AF855345

The Geo. S. Long Publication Series

Land Use and Forest Resources in a Changing Environment

The Urban/Forest Interface

EDITED BY

Gordon A. Bradley

UNIVERSITY OF WASHINGTON PRESS
Seattle and London

Printed in the United States of America

Library of Congress Catalog Card Number: 84-50977

ISBN O-295-96104-X cl
ISBN O-295-96145-7 pb

Proceedings of the National Symposium on the Urban/Forest Interface: Land Use and Forest Resources in a Changing Environment, held at the University of Washington, October 27-29, 1982, conducted by the Office of Continuing Education, College of Forest Resources, and sponsored by the University of Washington College of Forest Resources and the Geo. S. Long Fund.

Institute of Forest Resources Contribution No. 49.

The Geo. S. Long Publication Series

The Geo. S. Long Fund was established in 1975 to promote a better understanding of forestry, natural resources, and conservation. The endowment was made by Miss Helen Long of Tacoma to the University of Washington in memory of her father, a distinguished timber industry executive and natural resources conservationist, who opened the first Weyerhaeuser office in Tacoma in 1900 and served as the firm's manager, vice president, and chairman of the executive committee.

The generous financial support provides an income to the University of Washington to carry out the intent of the founder with provisions that allow the dean of the College of Forest Resources to develop scholarly programs that enlarge and enhance the contribution which the College makes to society.

CONTENTS

IV: PHYSICAL AND BIOLOGICAL DIMENSIONS

V: SOCIAL AND PSYCHOLOGICAL DIMENSIONS

VI: COURSES OF ACTION

PREFACE

The 1980s represent a period of transition for forestry and forest land planners. As an extensive land use, forestry is now increasingly confronted by conflict with alternative land uses. These conflicts have resulted in both the conversion of forest land to nonforestry uses and, less dramatically, the modification of forest management practices.

Because of the traditional assumption that timber production is incompatible with urbanization and other noncommodity uses of land, foresters have viewed the urban/forest interface with concern—often as a political battlefield representing the last stand against the 'loss' (conversion) of forest land. From this perspective, the conversion of land use is judged undesirable, hence subject to governmental actions to slow or prevent it. Others, including developers and, in certain instances, urban planners, have seen land conversion as desirable and representing 'higher and best use' of land. From this vantage point, an orderly conversion process should be promoted. Still other interests may be caught in the middle, perceiving the benefits of both commodity and noncommodity values of the forest as well as the desirability of controlled growth—the best of both worlds. The question that might be asked is 'Can forestry and urbanization coexist and be made more compatible?' If so, how can this be done? If not, what processes determine land use patterns, and how does a landowner respond to these processes? Finally, how does government use the available regulatory tools to influence the land uses in the public interest?

The many dimensions of land use include the legal-political and administrative framework of both public and private land use decisions. Economic values (monetary as well as intangible) also play a large role. The social values and psychological perceptions behind individual choices likewise influence land use decisions. Finally, the physical and biological capability of the land itself is fundamental to land use decisions.

A greater awareness of the role of each of these dimensions allows us to examine the urban/forest interface and to evaluate critically the scope and effectiveness of policies designed to influence land use decisions. Only by understanding these complexities can we produce viable solutions to the questions being raised by foresters and urban residents alike.

ORGANIZATION OF THE BOOK

The organization of this book follows the order of presentations at the Urban/Forest Interface Symposium held at the University of Washington in October 1982. There are six major sections. In part I, chapter 1 is an introduction and overview. General concepts are discussed that are elaborated in subsequent chapters. Chapter 2 provides a detailed examination of forest land use in an urban context. Attention is focused on the setting of the urban/forest interface, including land use, land markets, ownership, and recent trends.

Part II discusses the legal, political, and administrative framework within which interface decisions are made and examines the constitutional and legal basis for public and private action in the use of land and forest resources. Part III provides insight into the economic dimensions and focuses on land and resource valuation and factors influencing use and allocation. Part IV examines the physical and biological aspects and discusses the productive potential of land and forest resources and the implications of changing land use. Part V analyzes the social and psychological dimensions and looks specifically at the values and beliefs held by individuals and communities concerning land and forest resources.

Part VI discusses approaches to solving interface problems. A thorough examination is devoted to strategies to achieve public and private land use and forest resource goals. The final chapter is to a limited extent a summary of the preceding chapters, but, more important, it provides key observations regarding the complexities of the urban/forest interface and identifies areas where further research is needed.

The book reflects the organization of the symposium from which the chapters were developed. A speaker/discussant format was used, and therefore, except for the first and last chapter, the sections consist of the presentations by the principal speakers followed by those of their respective discussants.

ACKNOWLEDGMENTS

Many individuals contributed to the symposium and this book. Without the enthusiastic response and thoughtful papers prepared and delivered by the symposium speakers, this book would not be possible. To those individuals I am indebted.

The development and organization of the symposium and this book were guided by the efforts of a steering committee whose members included University of Washington Professors Gordon A. Bradley, Robert G. Lee, Donald Miller, and Thomas R. Waggener. Many other individuals from both the public and private sectors of forestry and planning also provided helpful suggestions on initial proposals for the symposium.

Support for the symposium and the publication of this book was provided by several sponsors and contributing organizations. Sponsors include the College of Forest Resources and the Department of Urban Planning at the University of Washington; the Society of American Foresters, Land Use Planning and Design Working Group; the American Planning Association, Washington Chapter; and Resources for the Future, Forest Economics and Policy Program. Contributing organizations and the individuals representing those organizations, who were kind enough to provide funding in a time of already strained budgets, include:

The Burlington Northern Foundation on behalf of B.N. Timberlands, Inc., Glacier Park Company, and Meridian Land and Mineral Company, represented by Donald K. North.

The Geo. S. Long Fund, administered by the College of Forest Resources, University of Washington, and represented by David B. Thorud.

The National Park Service, Pacific Northwest Region, Cooperative Park Studies Unit, University of Washington, represented by Donald Field and Darryll Johnson.

Weyerhaeuser Foundation, represented by Mary Hall.

Weyerhaeuser Real Estate Company, represented by James Nyberg.

The Washington Forest Protection Association, represented by Donald Chance.

It should be mentioned that Donald Chance, director of Land Use for the WFPA, also provided invaluable assistance in the development of the initial idea for the symposium. Through his efforts, a two-day field tour was conducted in the fall of 1981, during which foresters and planners were brought together to discuss conditions and potential problems associated with urban development and forestry on urban fringe lands near Seattle. During the field tour many more questions were raised than could be answered. Consequently, the Urban/Forest Interface Symposium was conceived.

Individuals who devoted countless hours to the mechanics of organizing the symposium include Reid Kenady, Mary Kenady, Ruth Jacobson, and Diane Nivaro from the Office of Continuing Education at the College of Forest Resources and Karin McDougall from the Management and Social Sciences Division of the College of Forest Resources. In addition, Polly Dyer of the Institute for Environmental Studies at the University of Washington and Marilyn Matthews of the Washington Forest Protection Association provided invaluable assistance in the area of media relations.

I am also grateful for the assistance provided by Kathi Grier who served as publications coordinator, Lucille Lewis who spent many hours at the word processor, and Leila Charbonneau for her editorial expertise.

Last, but certainly not least, was the ever dependable group of students from the College of Forest Resources who took care of all the details that distinguish successful programs from those that otherwise would not have been successful. This dedicated group included Stephanie Amoss-Anderson, William Frymire, Margaret Hartzell, Denver Hospodarsky, Gary Nelson, Debra Salazar, and Jessica Sall.

To all of these individuals and organizations, I offer my sincerest thanks and appreciation.

Gordon A. Bradley

Land Use and Forest Resources in a Changing Environment:
The Urban/Forest Interface

PART I

THE SETTING

1

THE URBAN/FOREST INTERFACE

Gordon A. Bradley

Attention has increasingly been drawn to problems associated with forest management near urban areas, or on the urban fringe. Much of the concern has been over urbanization of agricultural land. Numerous studies have explored the complex issues of agricultural land use, urban encroachment, and the effects of conversion of agricultural lands to nonagricultural uses. The urban/forest interface is a term chosen to draw attention to yet another area of concern regarding the use of land and natural resources on the urban fringe, in this case the relation between urban land use and forest resources.

Although several articles, books, and symposia have addressed various aspects of the problems of land use conversion in urbanizing areas (Clawson 1971, Healy and Short 1981, California Department of Forestry 1981, Waggener 1971, and Soil Conservation Society of America 1979), only recently has the subject been centrally addressed and similarly titled in the professional literature (Vaux 1982). If the professional literature has been limited in addressing the topic, the popular media have not. Reports of disputes between forest land managers, government planners, and citizen activists appear frequently in the daily press.

The concerns generally center on two issues: development that encroaches on forest lands, and the type of resources to be produced by forest lands and the methods of production.

In the case of urban encroachment, an issue raised frequently is the reduced supply of forest resources. Urban encroachment within commercial forest areas attracts the attention of the timber industry, which is interested in the potential ramifications to management costs and nuisance conflicts. Environmental interests that foresee losses in noncommodity resources of the forest are also concerned about forest land conversion. When questions other than the conversion of forest land arise, they usually involve issues of balancing commodity and noncommodity forest resources. While timber and minerals are extracted relatively easily from remote forest locations, their production becomes more difficult, and often a point of controversy, near urban areas, where urban interests may place higher values on amenities, recreation, and wildlife. More often than not, little distinction is made whether the lands are held in public ownership or by private interests.

In all cases, the question of what attribute is valued in land and forest resources is critical, as well as the questions of the appropriate criteria to be used in defining values, the process of choosing between values, and ultimately how resource values are distributed among potential recipients. This brings up fundamental questions that are addressed in this book. Can forestry and urbanization coexist? If so, how can this be done? If not, what processes determine land use patterns and how does a landowner respond? Finally, how does government use the available policy tools to 'manage' the land in the public interest? In order to address these questions, it is necessary to understand urban and forest resource uses as well as the motivations of all those involved in the allocation, ownership, and use of land on the urban/forest interface.

WHAT IS THE URBAN/FOREST INTERFACE?

The urban/forest interface, simply defined, is two traditional land uses occurring near or adjacent to one another—forestry and urban development. This is not always a problem, since many forestry and urban land uses are compatible and, in many situations, complementary. It is only when interface activities generate real or perceived negative effects that concern is expressed. This works both ways: forestry may affect the urban user in some ways, and effects may accrue to forestry as a result of urban uses. When concerns are expressed, conflict often follows, and the battlefield for conflict is more often than not a city council room, a county courthouse, or state and federal courtrooms and legislative chambers.

Although certain concerns, or conflicts on the interface, manifest themselves in spatial terms, they soon take on sociopolitical dimensions, and one set of values is pitted firmly against another. It is at this point that understanding the interface, its problems, and possible solutions becomes complex. So what is the urban/forest interface? As will be discussed in this book, it is sometimes thought of as a geographic phenomenon, but more often than not the root dimensions of the problem are sociopolitical. This perspective is important in understanding the interface and is addressed from several perspectives throughout this book.

THE PROBLEM

To think of the urban/forest interface as a contemporary phenomenon would be to overlook two centuries of settlement history in the United States. Obviously every farm, town, and metropolitan area was carved from either forested or open and wild landscapes (Clawson 1973). Settlements were established near natural water access points and were relatively compact. Over the years, development spread, overcoming the obstacles presented by wild lands—following canals, wagon trails, railroads, and eventually interstate highways. Early settlers viewed wild lands as either a barrier to be conquered or a major source for commodity production, but today's settlers often move to wild land areas with different attitudes and motivations. Although resource production at various levels may be a factor, the attraction of rural landscapes and the appeal of a home on a large lot strongly influence human decisions to move beyond established cities. In any event, our built environment, as we know it today, covers land that until settled was wild. Therefore, while some people may view the urban/forest interface as a recent phenomenon, it is in fact one of long standing and constant change.

Although it is not suggested that all shifts in land use are either good or inevitable, certainly the dynamic forces of nature, the economy, and political choice will continue to alter the appearance and use of land. In spite of these changes, major difficulties do arise by virtue of how society perceives change of any kind. As noted by Donald Schon in *Beyond the Stable State*(1971), 'A social system does not move smoothly from one state of its culture to another. In processes of social transformation, societies move from a relatively stable state through a zone of disruption to a new zone of relative stability. Something old must come apart in order for something new to come together. But for individuals within the system, there is no clear grasp of the next stable state—only a clear picture of the one to be lost.'

In many cases, our losses due to changing land use have been significant and irreversible. But many changes in land use have also resulted in benefits to society that far exceed their cost. Whether one is a forester, planner, or environmentalist, the new landscape envisioned and agents of change required to produce it may afford reasons for concern. Even when a vague image of the next state can be perceived, acceptance may be only partly forthcoming. Again, quoting Schon, 'Ideas are often slow to come into good currency, and once in good currency and institutionalized they are slow to fade away. By the time ideas have come into good currency, they often no longer accurately reflect the state of affairs.' This situation is not surprising when one considers a final observation by Schon, 'Belief in the stable state is belief in the unchangeability, the constancy of central aspects of our lives, or beliefs that we can attain such constancy. Belief in the stable state is strong and deep within us. We institutionalize it in every social domain. We do this in spite of our talk about change, our apparent acceptance of change and our approval of dynamism. Language about change is for the most part talk about very small change, trivial in relation to massive unquestioned stability.... Moreover, talk about change is as often as not a substitute for engaging in it.'

This is not to suggest that matters of concern regarding the urban/forest interface have not been addressed in an attentive fashion, but rather that much of our concern and resultant conflict may go much deeper than most participants imagine or would care to believe.

Therefore, while the urban/forest interface phenomenon may be an old one, it is important to recognize that (1) we have only recently begun to study and understand it in any detail, (2) we are finding that it is a phenomenon of considerable economic, sociopolitical, biological, and legal complexity, (3) any shift or change from what appears to be a traditional landscape or land use practice is likely to meet some resistance, if for no other reason than it runs contrary to popular beliefs and values, and (4) there are certain characteristics that may distinguish the interface of several decades ago from the one we are facing today.

Several factors are important with regard to how the urban/forest interface of the 1980s differs from earlier urban encroachments in wild land settings. At this time, land use decisions are guided by different laws and regulations and in many cases more complex ones than in the past. Certainly the many major environmental and land use laws of the late 1960s and early 1970s reflect society's attitude to the use of land and natural resources in general, and problems on the interface in particular. Many such laws have been national in scope. However, environmental, land use, and forest practice laws and regulations at the state level have probably been aimed to a greater extent at interface concerns. In addition, local jurisdictions, both cities and counties, have developed a variety of ordinances to deal with problems perceived to exist on the interface. Many of these ordinances are in the context of new or revised comprehensive land use plans. Involvement of local governments in land use activity has reached new levels. This is to be expected, since often they have the most at stake. They also frequently have the greatest power to address interface problems.

Our population has shifted substantially and continues to shift relative to the initial settlement patterns of this country. Although population was once compact, it subsequently dispersed and then reclustered in urban and industrial areas. Now people are again moving out of cities into remote rural locations. In many instances this outward movement may only be to the suburbs of major cities, but significant numbers are also going to rural, small-town communities and to the unincorporated countryside adjacent to urban centers.

The management considerations of land use and resource production, while still operating under the traditional theories guiding allocation of scarce resources, must now deal with dramatic changes in the market value of land and natural resources. Economic analysis must also attempt to factor many nonmarket values into land use and resource allocation models. With regard to values, our perception of forest land has become more complex and in certain sectors more intense. With legislation now reflecting these nonmarket values, there is greater political opportunity for interested parties to ensure that values considered important for society will be maintained. Therefore, while the urban/forest interface is in a constant state of change, the conditions under which change occurs and the rate of that change have varied significantly over the years. It was, of course, the purpose of the Urban/Forest Interface Symposium to explore an old phenomenon rather comprehensively in light of the conditions of the 1980s. By doing so, we may better appreciate the dynamics of change and locate ourselves within the current zone of transformation that the urban/forest interface represents. This is at least a necessary first step in the process of exploring alternative solutions to present-day conflicts—solutions that are sufficiently flexible, effective, and acceptable in addressing tomorrow's problems.

THE URBAN/FOREST INTERFACE TODAY

The present zone of transformation represented as the urban/forest interface can be better understood by discussing the competing uses for urban fringe lands and the principal motivations of landowners. This also provides insight into the complexity of the problem and why a search for immediate and lasting solutions is a major challenge for foresters and planners.

Figure 1 characterizes the typical zones that exist on the urban/forest interface. The continuum, though it suggests spatial importance, recognizes the political and social nature of the problem. The zones, as arrayed along the continuum, suggest a decreasing degree of urban encroachment as one moves from left to right. This is, of course, not always the case, and in many instances the zones are found in a variety of combinations. The basic purpose of the continuum is to identify graphically the primary interface uses, the incentives to conduct such uses, and potential problems associated with a landowner's choice in pursuing a particular land use goal.

The five zones include development, conversion, modified practice, nonindustrial forestry (NIF) and parcelization, and forest. To the left of the development zone are typical built-up city environments and to right of the forest zone are remote wild land settings that are either inaccessible for the most part or lands designated for preservation. Preservation lands would include national parks, wilderness areas, and land set aside for other scientific or amenity purposes. Thus the interface zones are in many instances, as Daniel Jack Chasen (1977) uses the term, 'up for grabs.'

Development Zone

The development zone—that is, residential development in a forest environment—is no doubt the popular image of the urban/forest interface. Of course many other types of development occur in this zone, including commercial, recreational, and industrial activities. This zone may be attractive for development for several reasons, not the least of which is the potential of lower market prices for land as one moves away from urban areas. In many cases, these less developed surroundings still within easy commuting distance to cities provide many amenities not found in built-up areas. Open space, trees, streams, and lakes are a few of the attractions that may lead developers and home buyers to the development zone. Assuming that these developments are carefully planned, and in accordance with local capital improvement and other land use programs, many of the amenities sought can in fact be realized with few problems.

Severe problems may occur when the developers, or unknowing buyers, discover that insufficient buffers have been provided as part of the development. Suddenly, the amenities thought to be a permanent part of the landscape are in fact a component of another public or private

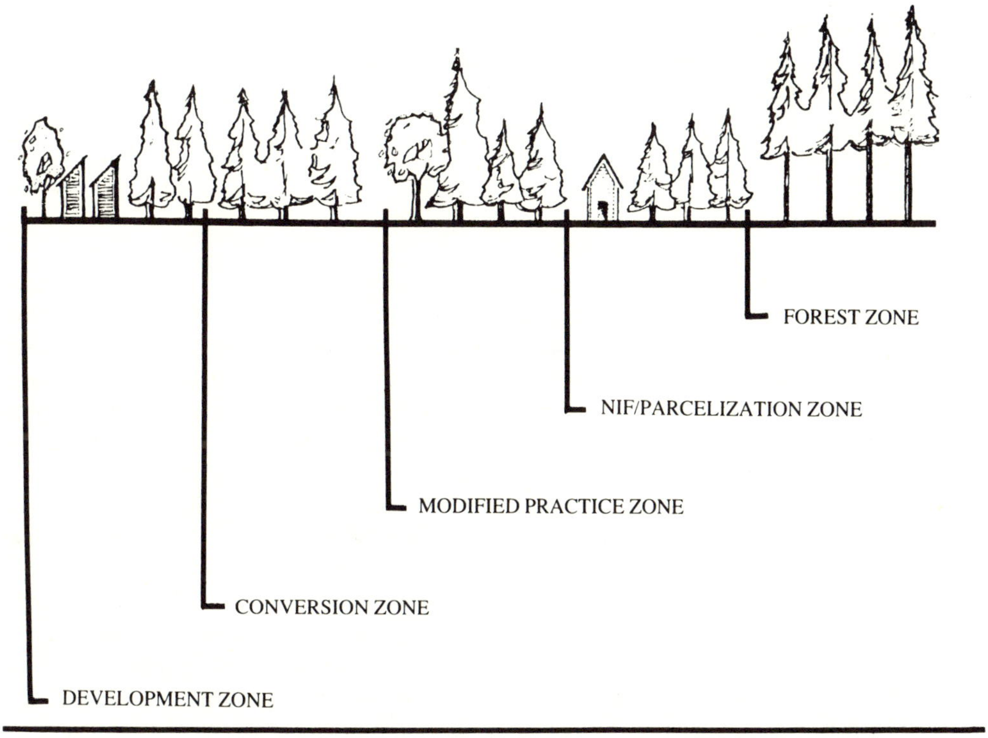

Figure 1. Urban/forest interface continuum.

owner's land and timber management plan. Not only does the landscape take on a new image after harvesting or development, but many forestry development practices (including the use of fire, chemicals, and heavy equipment) are in conflict with the new owner's perceptions of how adjacent lands should be used.

Also, the availability and cost of public services may not be what landowners in the development zone expected. If public services are made available at a level and cost comparable to urban areas, they may in fact be provided at the expense of other landowners in the taxing jurisdiction. Such tax-transfer situations raise further issues. While the provision of public services is of concern to the new landowner, it is probably a greater concern to financially strapped local governments, which in many cases are ill-equipped to maintain the existing infrastructure, let alone expand into new areas. And long-term residents who have grown accustomed to life without public services are sometimes reluctant to be assessed for services they do not desire or require.

Another problem pertaining to the development one is that of managing a forest environment amid development. A forest stand is in a constant state of change, or biological succession. To expect it to remain unchanged once development occurs, and to fail to anticipate the effects of development on the management of the forest, is to court disaster. Instances of trees blowing down on structures due to improper thinning and soil disturbance are all too common, as are problems with vegetation dying from overwatering, improper fertilization, and air pollution. Also, to suppress the natural forces of nature, such as the periodic occurrence of fire, without other mitigating measures, may result in the type of fires all too familiar to California's chaparral region.

The development zone is obviously attractive for many reasons. Many of these attractions can be maintained through careful analysis prior to development. When public services, adjacent land use, and forest management needs are ignored, however, the development zone becomes a potential problem of considerable magnitude.

Conversion Zone

The conversion zone is typically thought of as land in forest cover that is expected to change to another use fairly soon. This may be productive forest resource land, or it may be marginal land in terms of its ability to produce forest resources—outputs including recreation, timber, and wildlife. Conversion is usually prompted by an owner's expectation that a more profitable use exists for the land. This use may be a form of urban development or more intensive forms of natural resource development, including agriculture. Several factors may contribute to a landowner's decision to convert the land use, including favorable zoning decisions by local government, a change in tax status, or a change in the use of adjoining lands. While a decision to change use may be made on the spur of the moment, it may well have been made as much as twenty years earlier (Brown et al. 1981).

Regardless of the motivation and intentions of the landowner, the mere mention of land conversion raises numerous issues and emotional debate. Local government and development interests may see opportunities for increased revenues through higher taxes. On the other hand, local government along with environmental interests may view land conversion as creating additional financial burdens and diminishing an already scarce natural resource base near populated areas. Others, including the timber industry, question conversion on the grounds that it diminishes the timber supply and may create problems in managing lands for timber purposes. In some cases it is the development subsidiaries of timber companies that are converting forest land to other uses, including residential development and agriculture (Healy and Short 1981).

While there are obvious advantages and disadvantages associated with forest land conversion, certain management problems facing the landowner are often overlooked. Once a decision is made to change uses—however distant in the future—the landowner should manage the land consistent with his new land use objectives and with an eye toward maximizing whatever benefits are desired. An excellent example would be a timber landowner who is planning to convert commercial forest land to residential use ten years hence. To ignore possible financial gain from the land for a decade would not be prudent from an investment perspective. Yet, to continue conventional forest practices with the associated levels of investment might also be unwise. This might call for new forest practices and investment strategies that could be termed 'conversion practices.' Such practices might include modified harvest methods or reduced investment in fertilization or thinning when benefits might not be realized within ten years. Also, leaving certain species (or even individual trees) standing, even though they have immediate commercial value, may in fact contribute greater benefit to the future proposed use, including an increase in land value. Road layouts for harvesting might also be modified to fit the future land use scheme. Rather than investing in grading, drainage systems, and ballast for harvest roads, and then having to overcome problems associated with these roads in future residential building and landscaping, it could be to the advantage of the forest landowner to design road systems for harvesting that can be used at least in part in the new development. Finally, assuming that the timber landowner may be interested in continuing to manage lands for forestry adjacent to the new development, special attention should be paid to the layout and density of the development.

Many common conflicts between forest practices and urban uses, including residential development, can be avoided through innovative design that includes setbacks that cluster uses at greater densities, vegetative buffers, structural orientation, and predevelopment stand manipulation.

Like the development zone, the conversion zone may provide a logical and noncontroversial transition of the landscape from one use to another. If such uses are inconsistent with jurisdictional needs, individual perceptions of resource values, or prudent land practices, forest land conversion may continue to be a point of major controversy.

Modified Practice Zone

The modified practice zone is, as the name implies, an area in which traditional forest land management practices are modified to allow for the continuation of the present forest uses, while at the same time making that use more compatible with adjacent land uses. Presumably, the modified practices employed are not so costly as to make management impractical from an investment perspective. Otherwise, the landowner would begin to employ 'conversion practices' with an eye to a more profitable use.

A landowner's motives for modifying land management practices may be altruistic, an attempt to avoid costly litigation, or the result of government regulation. In any case, the modified practices are intended to maintain or increase certain benefits while allowing for an acceptable flow of the resource benefits that are the main concern to the landowner.

For example, a timber landowner whose main objective is timber production may find it necessary to modify silvicultural practices to continue producing wood and fiber. Modified practices may include changes in harvest size and location, the use of shelterwood and partial cuts, and leaving buffers in certain areas. Also, road layouts and harvesting systems may be subject to change, as would the use of fire and chemicals for the traditional purposes of brush control land regeneration.

Modification, however motivated, presumably allows for the maintenance of aesthetic values, wildlife habitat, and air and water quality. In certain cases, modifications may also be undertaken for health and safety reasons.

For modified practices to be effective, it is obvious that a decline in primary resource production may occur. While this may represent a new and previously ignored cost of doing business, it is one that the landowner must accept if forestry use is to be maintained. Incentives or compensation programs may be called for if the public deems the maintenance of forestry activities desirable. Depending on the extent and origin of the modified practices required, it may be appropriate to provide the landowner with preferential tax advantages, direct subsidies, partial purchase of land use rights, or (in extreme cases) the outright purchase of the land in fee simple.

The modification of forest practices in the modified practice zone may be thought of as the inverse of the requirements of a landowner in the development zone. That is, in order to pursue a particular land use or forestry objective, the responsibility for innovative practices and design falls partly on the landowner in the modified practice zone. Of course, to expect landowners to assume full responsibility for making their use compatible with others, without some incentive, would be impractical both economically and politically.

NIF/Parcelization Zone

The NIF/parcelization zone is distinguished from others by virtue of parcel size. Two types of land use, thus two kinds of owners with different aims, are involved. One landowner has a parcel ranging from 5 to 500 acres; his main objective is to grow trees on this 'nonindustrial forest' (NIF). The other is the 'large lot' landowner whose land use objectives may vary, but who has purchased a lot in a zone where the local jurisdiction has restricted minimum lot sizes. Typically such restrictions are for lots of 5, 10, 20, or 40 acres. In some cases lot minimums may be as much as 140 to 640 acres. Often the landowner is seeking an affordable place to live, or a place to escape on weekends where the amenities of the forest may be enjoyed.

In NIF land uses, fewer problems generally exist. Management practices are less intensive and on a smaller scale than usually found on industrial forest lands, resulting in fewer conflicts with adjacent landowners. Where practices do pose potential problems, the NIF landowner would presumably be subject to the same requirements of an industrial forest landowner in the modified practice zone.

NIF lands serve several functions that are of importance to planners and foresters, as well as the NIF landowner. To the planner, NIF lands help maintain rural character and open space needs while at the same time producing commodity resources and presumably making a contribution to the local economy. To the forester, NIF lands are often a partial solution to potential timber supply problems. Approximately half of the available commercial timber in the United States, excluding Alaska, is held in small NIF landownership. To the NIF landowner, a small parcel may satisfy the 'American dream' of owning land, while at the same time providing either a supplementary or total income.

The extent to which a NIF landowner is motivated to manage a small parcel for resource production is a function of several factors. Knowledge about and an interest in practicing forestry are an obvious prerequisite which the landowner may already have, or may acquire over time through exposure to public and private NIF incentive programs. Knowledge and interest in practicing forestry are certainly necessary but are not sufficient for a landowner to achieve NIF land management objectives. Economies of scale and site productivity may also determine whether NIF land can be managed effectively and profitably. Small, low-site parcels are in most cases insufficient to provide an income of any consequence over time. Clark Row, in his research, found that parcel size is a major factor determining whether a landowner can manage a forest tract and expect to make a profit (Row 1978). This presents major problems when the local government establishes minimum lot sizes and assumes that such lots will be managed for resource production, but in fact it is not economical to do so because of the small size. This can result in any number of problems, not the least of which is idle land when the planning objective may be to make the land more productive.

Where resource production may not be the explicit purpose for establishing minimum lot sizes, or the principal motivation of the landowner, the phenomenon of parcelization exists. In this case, the land may be used for a primary or second residence or simply for investment or speculation. While the ownership of a large lot may be satisfying to the landowner, a variety of problems may exist. Without proper management, the productivity of the land for a variety of purposes may decline. Not only timber resource production, but fisheries, wildlife, and aesthetic resources may also decline. In addition, while the local government may have established minimum lot sizes to maintain the rural character or preserve certain environments, the costs in many cases outweigh the gains.

These costs take the form of attempting to supply public services wherever demanded, thus increasing the number of points of possible conflict with traditional resource uses and increasing the probability of health and safety problems where minimum road, building, fire, water, and sewer requirements are not established. Although the concept of large-lot zoning has been extremely popular with local governments for the past decade as a land use planning tool, the problems of parcelization are now beginning to surface in sufficient number and magnitude to raise questions about its effectiveness. Unfortunately, as one looks at a land use map of lands adjoining most metropolitan areas, the extent to which lands have already been divided into small parcels is staggering. Attempts to address this situation in any effective manner are formidable.

Forest Zone

The forest zone consists of land whose principal management function is to produce natural resources, including wood and fiber, recreation, wildlife, fisheries, water, and forage. This land is held publicly or privately, usually in large parcels. Because of its greater distance from urban development, fewer problems exist with regard to land use conflicts or the possibility of changing use in the foreseeable future. This is not to suggest that the forest zone is without problems. For both public and private lands, there are numerous legislative guidelines from all levels of government on how the resources will be managed. In most cases, these guidelines are intended to mitigate conflicts between forest resource uses rather than address problems associated with conflicts between forestry and urbanization. While these lands are sufficiently remote to be considered a forest zone, certain interests may see them in terms of eventual reclassification to other, less productive natural resource or development uses. By knowing the opportunities and constraints associated with the use of lands in the other zones, one can make better decisions regarding forest zone lands whereby negative consequences are minimized and greater public and private benefits are achieved.

PERCEPTIONS OF THE URBAN/FOREST INTERFACE

While the urban/forest interface continuum suggests a clear distinction between land uses and their associated opportunities and constraints, it is important to point out that in practice not all people have the same perceptions of the interface. Individuals are motivated to own or use land for different reasons. Use decisions are a function of the owner/manager's perceptions of the utility of this land. These variations in perceived utility lead to yet another, more fundamental problem: agreement on a definition of the urban/forest interface. Without some notion of where the problems lie and an understanding of their dimensions, effective solutions are unlikely.

As discussed for each zone, the issues of economics, aesthetics, physical and biological productivity, existing use, and jurisdictional boundaries are identified as motivating factors. When we think of each of these factors as criteria by which we define the interface, or glasses through which we view it, it becomes apparent why there is a lack of consensus regarding those lands that constitute the urban/forest interface. Without this consensus, it is also apparent why a concise definition of the problem has not been achieved and why certain proposed 'solutions' may miss the mark. As figure 2 illustrates, when using economic criteria, the interface between forest uses and urbanization is that line where the marginal value of land for urbanization is approximately equal to the marginal value of land for forest uses.

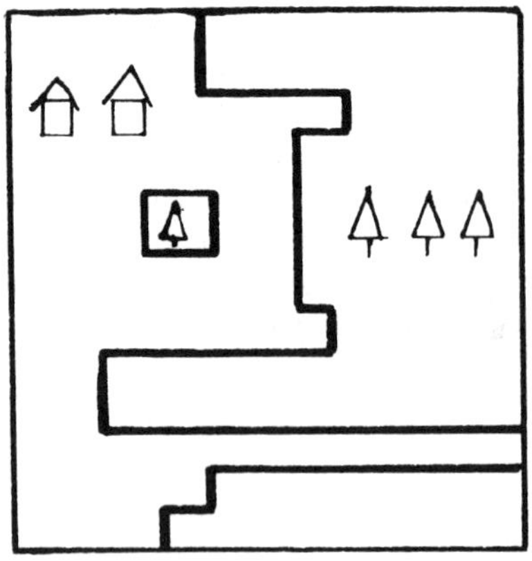

Figure 2. Economic criteria.

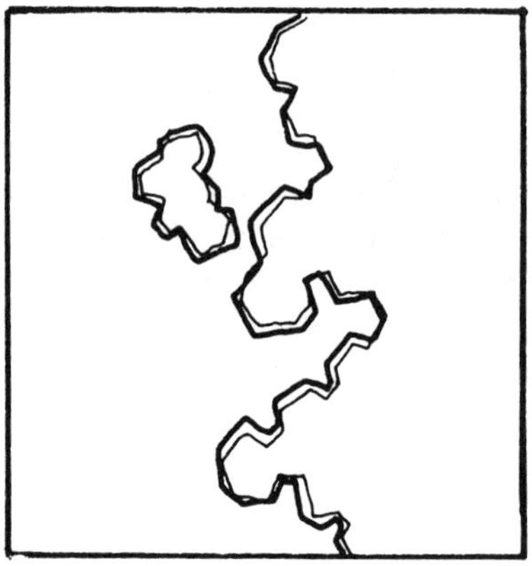

Figure 3. Aesthetic criteria.

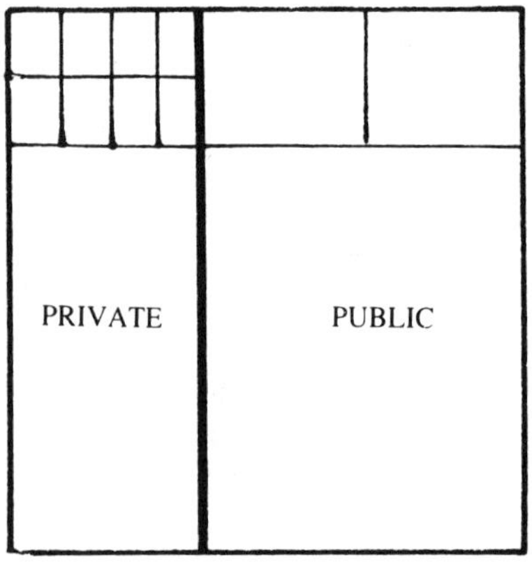

Figure 4. Jurisdictional parcel-size criteria.

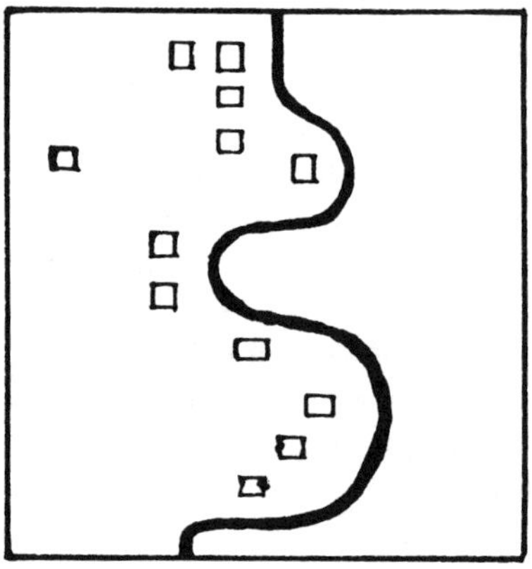

Figure 5. Development criteria.

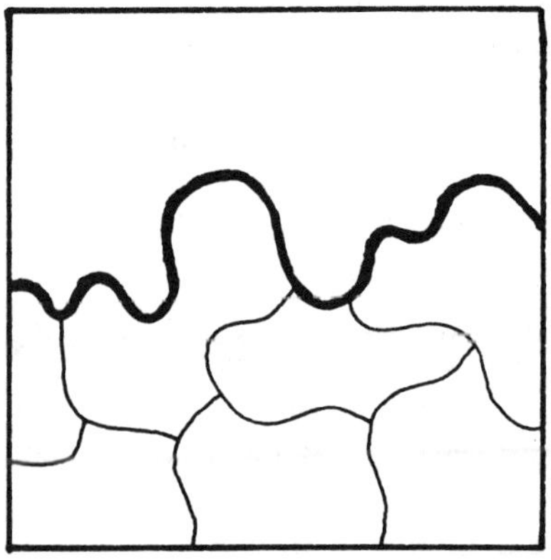

Figure 6. Physical and biological growth criteria.

From an aesthetic perspective (figure 3), the urban/forest interface is a wide and sometimes fuzzy boundary distinguished for the most part by the presence or absence of trees.

From a jurisdictional perspective (figure 4) the interface is sometimes thought of as either the boundary between public and private land or where parcel size begins to change from relatively small to relatively large lots.

Where structures appear on the landscape, the urban/forest interface may be defined by a line drawn between the built and unbuilt environments (figure 5). And finally, using physical and biological growth criteria (figure 6), the interface may resemble a soils map, where highly productive lands are distinguished from less productive lands. Obviously, to come to terms with the 'appropriate uses' for interface lands and the mechanisms whereby those uses are achieved without major conflict, all relevant factors and perceptions must be considered and made an explicit part of the strategies and policies that are implemented in order to achieve land use and forest resource goals.

TOWARD SOLUTIONS

Although strategies to achieve public and private goals in land use and forestry are discussed in detail later in the book, it may be useful to consider the idea of solutions on a conceptual level.

If our problems were simple, and if we were pursuing a single, clearly defined goal, our choices would be more straightforward. But, as the urban/forest interface continuum suggests, what we have is an ill-defined problem, or at best a problem with multiple definitions. Also, there are many actors who are simultaneously pursuing multiple goals. For some, the free market is the obvious solution. For others, governmental regulation is the only answer. Others may find combinations of these as yet untried approaches to land use and forest resource problems necessary. In all likelihood, given the setting in which interface problems occur and the rules by which society operates, the latter approach which combines strategies and relies on invention will probably prevail. For a program to be politically viable as well as effective in addressing interface problems, a variety of techniques is necessary.

The means are many and they can be classified in several ways. Work by Maynard Hufschmidt (1971), while focusing on means for achieving environmental quality goals, provides a useful framework for looking at potential solutions to interface problems.

According to Hufschmidt, means can be classified in many ways: short-term versus long-term, centralized versus decentralized, direct versus indirect, and legal-political versus market. The techniques, or levels of intervention discussed by Hufschmidt, include those commonly used by government as well as some that have not been applied extensively. Figure 7 illustrates the means of achieving environmental and land use goals along a continuum from indirect to direct intervention by government in decision-making and market processes. As noted by Hufschmidt, 'We would expect the effectiveness of direct intervention to be greatest in achieving specific short-run environmental quality goals, but to decline significantly in the long run. Conversely, indirect means are likely to have little payoff in the short run but will become increasingly effective in the long run and provide the opportunity to achieve a balanced contribution to major public policy goals. In addition, the political feasibility of indirect means is relatively high, while direct penalties on private activities, and some public activities, tend to be controversial and to conflict with other public goals. Given these assumptions, a strategy of mixing direct and indirect means appears preferable for achieving both immediate improvement and sustained maintenance of environmental quality' (p. 237).

INDIRECT

1. Environmental monitoring
2. Public information
3. Environmental research
4. Technology development
5. Legal action by citizens
6. Rewards to consumers and producers
7. Direct public investment and management
8. Modifying other government programs
9. Taxes and charges
10. Direct government controls

DIRECT

Figure 7. Check list of means for achieving environmental quality and forest land use goals along a continuum from indirect to direct intervention by government in decision-making and market processes (source: Hufschmidt 1971).

Without discussing here the various strategies in detail, it is useful to note how these techniques have been applied to the urban/forest interface zones. Figure 8 illustrates generally the relation between the use of the levels of intervention or techniques and the zones to which they are most often applied. While in certain cases there may be considerable use of a technique in a particular zone by certain organizations or governments, it is reasonable to assume that the relation between techniques and zones as depicted in figure 8 is generally reflective of where and how society has chosen to address its problems to date.

As indicated by the matrix, current efforts emphasize more direct levels of intervention. Where less direct approaches are used, they are unevenly distributed across the interface zones. If, as according to Hufschmidt (as well as Berry and Plaut 1978), the more direct the intervention, the less likely we can expect long-term solutions to environmental quality and land use problems, then we have been investing heavily in approaches that may not sustain themselves. It would appear that greater attention can and should be focused on long-term approaches across all interface zones, such that we understand the urban/forest interface in more substantive terms and presumably develop more effective solutions to problems.

Certainly that was the purpose of the Urban/Forest Interface Symposium and the intended hope of this book: to focus greater attention on a phenomenon that is not well understood, so that we may begin to make more informed choices about the use of land and forest resources on the urban fringe.

	Development	Conversion	Modified practice	NIF/Parcelization	Forest
Environmental monitoring		X			X
Public information					X
Environmental research					X
Technology development					X
Legal action by citizens	X	X	X	X	X
Rewards to consumers and producers				X	X
Direct public investment and management	X	X		X	X
Modifying other government programs	X	X	X	X	X
Taxes and charges	X	X		X	X
Direct government controls	X	X	X	X	X

X = Application of technique.

Figure 8. Techniques and their application to urban/forest interface zones.

REFERENCES

Berry, D., and T. Plaut. 1978. Retaining agricultural activities under urban pressure: A review of land use conflicts and policies. Policy Sciences 9(2):153-178.

Brown, H. J., R. S. Phillips, and N. A. Roberts. 1981. Land markets at the urban fringe. Journal of the American Planning Association 47(2):131-144.

California Department of Forestry. 1981. Timberland conversion study. Sacramento, California.

Chasen, D. J. 1977. Up for grabs, inquiries into who wants what. Madrona Publishers, Seattle.

Clawson, M. 1971. Suburban land conversion in the United States: An economic and governmental process. Published for Resources for the Future, Johns Hopkins Press, Baltimore.

———. 1973. Historical overview of land-use planning in the United States. In: D. M. McAllister, ed., Environment: A new focus for land use planning, pp. 23-54. The National Science Foundation, Washington, D.C.

Healy, R. G., and J. L. Short. 1981. The market for rural land: Trends, issues, policies. The Conservation Foundation, Washington, D.C.

Hufschmidt, M. M. 1971. Environmental quality as a policy and planning objective. Journal of the American Institute of Planners 37(4):231-242.

Row, C. 1978. Economics of tract size in timber growing. Journal of Forestry 76(9):576-582.

Schon, D. A. 1971. Beyond the stable state. W. W. Norton and Company, New York.

Soil Conservation Society of America. 1979. The diminishing forestland base and land use planning in Washington. A conference sponsored by the Olympic View Chapter of the Soil Conservation Society of America, Sept. 28.

Vaux, H. J. 1982. Forestry's hotseat: The urban/forest interface. American Forests 88(5):37, 44-46.

Waggener, T. R. 1971. Economic factors determining land use allocation. Paper presented to the annual meeting, Society of American Foresters, Inland Empire Section, Spokane, Washington.

2

FORESTS IN AN URBAN CIVILIZATION: LAND USE, LAND MARKETS, OWNERSHIP, AND RECENT TRENDS

Robert G. Healy

America's forests are firmly embedded in the context of an 'urban civilization.' To define this context is to begin to perceive an important problem for all those concerned with the use and conservation of the forest resource.

What do I mean by an 'urban civilization'? It means simply that about three-quarters of the nation's population lives in urban areas. This is quite novel if one takes a historical view: as recently as 1910 the majority of Americans lived in rural areas, mostly on farms. But the urbanization process in a purely demographic sense is slowing down, not speeding up. The 1980 census revealed that for the first time in nearly a century the rural population has been growing almost as rapidly as the urban population. And during the 1970s, the population of nonmetropolitan areas—America's small towns and countryside—grew significantly faster than that of its great metropolitan centers. We must in fact take a broader view of urbanism if we wish to understand the urban stresses that modern American society places on its forests.

Sociologist Irving Kristol (1972) has written: 'What the textbooks call the "urbanization of America" is not at all the "citification of America."' Kristol suggests that the United States has become an urban civilization not just in its geographic concentration of population but in the sense that its predominant culture is homogenized and cosmopolitan, rather than provincial. The characteristics of this urban civilization most relevant to forests are that the great bulk of its citizens do not make a living from the soil, they have high average incomes and levels of education, and they have a high demand for leisure pursuits and for a wide variety of amenities. This urban civilization is also characterized by certain kinds of industrial, commercial, and residential activities that have historically been found mainly in large cities, although they are increasingly spread throughout the national landscape.

It is the contention of this paper that an urban civilization makes new and difficult demands on its forests, even as it imposes constraints on forest land and on forest managers. These impacts are of two principal types. First, the urban civilization looks to forest land as a convenient site for many of its activities—for its factories, its highways, its reservoirs, its stores and homes. In this way, forest land is 'urbanized' or 'converted to urban use,' although this does not necessarily mean that it becomes part of a contiguous built-up area. This chapter will analyze this competition for land and draw some comparisons with the much better known problem of the urbanization of cropland. Second, urban tastes and values influence how and for what purposes that land remaining in forest use will be managed. American forests now have a large urban constituency. Historian Samuel P. Hays has noted the increased interest taken in forests by people who neither own them nor rely on them for fuel or fiber. He observes: 'After World War II, discovery of the nation's wildlands by an increasing number of people led to an intensity of human involvement in the forested environment unknown earlier. People discovered natural lands and came to know them personally by hiking through them, observing them closely, photographing them.... Through such personal involvement these areas came to be thought of as one's own "turf" to be known, enjoyed, described, communicated and protected' (Hays 1980).

This protection of turf has led not only to greater involvement by citizens and interest groups in the management of government-owned forest lands but also to public demands for a say in how private lands are managed. One result has been an increase in legislation regulating forest practices, and a substantial body of pollution and land use laws that affect the use of forest land.

The urbanite's love affair with the American forest is a subject with complex psychological roots, many classic manifestations in literature and art, and quite varied political results. I mention it here to acknowledge its importance, and to admit that its analysis is quite beyond my disciplinary competence. I intend rather to concentrate on the *physical and economic impacts* of the urban civilizaton on forest land. The most obvious of these impacts is land use change.

LAND USE CHANGE

During the last two or three years, a tremendous amount of attention has been paid to the conversion of cropland to urban use and to the attendant danger to future food supply. The farmland conversion problem has been the subject of a federal interagency study (U.S. National Agricultural Lands Study 1981), several state and regional reports (MSM Regional Council 1980, People for Open Space 1980), conferences, and an avalanche of books and articles (Crosson 1982, Woodruff 1980, Sampson 1982). No similar concern has been voiced about the loss of forest land, save in a few publications read only by forestry professionals. It is interesting to compare the magnitude of recent crop and forest land changes.

According to the National Agricultural Lands Study (NALS 1981), between 1967 and 1975, 5.4 million acres of cropland, about 675,000 acres per year, were converted to urban, built-up, transportation, or water uses. This was a loss of 1.3 percent of the cropland base in 1967. During the same eight-year period, however, the U.S. lost 6.5 million acres of forest to those same uses. This was a rate of over 800,000 acres per year, and amounted to 1.5 percent of the private forest land base. (Consideration of the extensive publicly held forest lands, however, would reduce the proportion converted to 1.0 percent.)

Table 1 reveals significant regional variations in the amount and rate of conversion. The greatest number of acres lost was in the Southeast, followed by the Northeast and Appalachian regions. In terms of percentage of forest land converted, the same ranking holds, with the Southeast losing 2.6 percent of its forest land over the eight-year period, against 2.1 percent for the Northeast and 2.0 for Appalachia. Conversion of forest land to water areas accounted for approximately one-third of the conversion. Presumably, this represents forested stream valleys and flood plains inundated as a result of reservoir construction.

At present, considerable controversy swirls around the NALS. Some participants in the fray have attacked the quality of the NALS conversion data, which originated in a 1977 inventory (U.S. Soil Conservation Service 1977) by the Soil Conservation Service, U.S. Department of Agriculture (SCS). Questions about the data range from accusations that the actual amount of conversion is overstated (Simon 1982) to observations that the 1967-75 period may have been a period of unusually rapid land use change (Raup 1980). Other critics believe that even if the NALS/SCS data are accurate, the foreseeable rate of loss is not high enough to pose a serious threat to U.S. agriculture (Crosson 1982). One might ask the same questions about forest land conversion: Do available data accurately measure the amount of conversion? Are past trends likely to continue? Will urbanization interfere with the traditional uses of forests?

Table 1. Changes in Forest Land Use by Region (1967-75).

Region	Conversion to Urban Areas	Conversion to Water Areas	Total Conversion
Northeast	1,026,275	401,132	1,427,407
Lake States	191,960	70,135	262,095
Corn Belt	279,890	151,910	431,800
Northern Plains	27,117	26,479	53,596
Appalachia	1,086,895	210,722	1,297,617
Southeast	938,402	984,897	1,923,299
Delta	273,270	51,531	324,801
Southern Plains	171,568	227,499	399,067
Mountain	18,264	27,900	46,164
Pacific	330,647	0	330,647
Total	4,344,288	2,152,205	6,496,493

Source: NALS 1981.

Before looking more closely at the relation between forest and urban uses, however, we should consider an important difference between forest land and cropland. Forests tend to be a residual use of rural land: trees occupy land that is not currently in much demand for growing crops or providing pasturage. If the demand for cropland falls, marginal cropland will be abandoned to forest; if demand for cropland rises, forests will be cleared away. Of course, some forest land is simply too steep or dry or infertile to be physically suitable for crop or pasture use. But much of our best forest land is in some sense in constant, though sometimes not obviously apparent, competition with these other uses. This substitutability has been most dramatically illustrated in New England, where the landscape has changed in the last two hundred years from primeval forest to mainly cropland and pasture and back to forest again.

Historically, the relative prices of food and wood have been such that forestry has found it difficult to compete for land that farmers want. Table 2 shows the prices that alternative uses currently command for the exclusive use of land. The competition creates a sort of economic hierarchy—a hierarchy in which forests rank near the bottom. There is no immutable economic law perpetuating this ranking. Yet it is clear that for forests to compete for fertile land against crops or even some intensively grazed livestock, the relative prices of forest products would have to be far higher than they are today.

The existence of this hierarchy means that forests are affected by urbanization not only directly but also indirectly. If cropland is urbanized, it will frequently be 'replaced' by converting forest land to crops. But if technological change in agriculture quickens, or if crop exports fall, the lesser demand for cropland will make more land available for growing trees.

The idea of the land use hierarchy is implicit in the statement that the U.S. has 127 million acres of 'potential cropland'—land suitable for growing crops but not now in crop use (NALS 1981). Most of the potential cropland is not unused, but is in uses lower in the hierarchy. Nearly

Table 2. Land Use Hierarchy.

Use	Typical Land Value (per acre)
Urban uses*	$5,000-50,000+
Orchards, specialty crops	2,000-7,000
Rural residences	1,000-5,000
Corn, soybeans	1,000-4,000
Developed pasture	300-1,000
Wheat, small grains, hay	200-1,200
Forestry (bare-land value)	100-600
Range-land grazing	50-200

*Urban uses include various intensive land uses, including housing subdivisions, industry, and commercial establishments.

31 million acres are now classified as forest land. Thus, if each acre of cropland urbanized is 'replaced,' an average of one-quarter acre of forest land will be converted to crops. (In fact, given differences in fertility, the ratio of forest land converted per acre of cropland replaced should be somewhat higher than this.) Moreover, about 40 percent of the potential cropland is now in pasture. If this land is converted to crops, either to replace lost cropland or simply to increase crop output, some forest land will likely be converted to pasture.

The NALS/SCS data indicate that conversion of forest land to crops has been an important mechanism by which the stock of agricultural land has been maintained in the face of urban pressures. During the same time that 5.4 million acres of cropland were being urbanized or inundated, fully 10.9 million acres were converted from forest to crops. An even larger amount of forest was converted to pasture or range, again more than offsetting the loss of grazing land to urban use.[1]

These replacement demands tend to be unappreciated, for it is virtually impossible to trace the chain of cause and effect. An acre of Illinois soybean land, lost to a housing development, may be 'replaced' by the clearing of bottom land hardwoods in Louisiana. Or the building of a reservoir that floods Tennessee farmland may eventually result in the creation of a new pasture in eastern Oklahoma.

It should be noted that net shifts between categories are the product of much larger gross shifts. Even as some forest land is being cleared for crops, other cropland is being abandoned and reverts to forest. Although NALS/SCS data showed net losses of forest land with respect to

1. NALS/SCS data indicate that 62.3 million acres of forest land were moved into the pasture or range category during the 1967-75 period. This undoubtedly reflects not only actual land clearing but the reclassification of low productivity forest lands (especially in the Mountain and Southern Plains regions, where 37.9 million acres shifted categories) as grazing land. Nevertheless, it seems clear from the regional data, as well as from independent reports of land clearing for pasture (see text), that the 4.3 million acres of pasture urbanized were more than offset by conversion of forest land to pasture. The amount of forest land cleared depends not only on the relative price of crops, cattle, and wood but also on the cost of clearing, estimated in one study at about $300 per acre (Herrington and Shulstad 1982).

both crop and pasture uses, the data also indicated that during the period under study, 8.3 million acres of cropland reverted to forest, as did 14 million acres of pasture and range. In fact, such was the magnitude of these gross changes that 21 percent of the land classified as forest in 1967 was no longer forest in 1975, while 7 percent of the land classified as forest in 1975 had not been forest in 1967. (Again, it is impossible to separate physical land conversions from reclassifications due only to redefinition.)

Although the NALS/SCS figures offer the only national source of information on land conversion, a host of regional and small area studies offer additional evidence. Most are studies by the U.S. Forest Service regional offices on the status of 'commercial forest land.' (Commercial forest land is defined as land capable of producing crops of industrial wood at the rate of at least 20 cubic feet per acre per year and not withdrawn from timber production.) They, too, generally support the contention that forest land is being urbanized, as well as being converted to agricultural use. Table 3 summarizes conversion pressures thought to be important in major forest regions.

Table 3. Major Reasons for Loss of Commercial Forest Land.

South:	Clearance for crops (MS, AR, LA, GA, NC, FL)
	Urbanization (AR, MS, LA, TX, GA, NC, FL)
	Clearance for pasture (AL, AR, MS, TX, OK, GA, NC, FL)
	Utility rights of way (LA, GA, NC)
	Water impoundments (TX, GA, NC, FL, KY)
	Reservation for parks or wilderness (GA)
	Roads and highways (GA, NC)
	Reclassification to noncommercial forest (GA, NC, FL)
North:	Urbanization (MN, MD, MA, CT, MI, MO, PA)
	Clearance for crops (MN, MO, PA, DE)
	Clearance for pasture (MO, MN)
	Rights of way (MO, MI, PA)
	Mining and industrial (PA)
	Water impoundments (PA)
	Reservation for parks or wilderness (MN, MO, MI)
	Reclassification to noncommercial forest (NH, PA)
West:	Clearance for crops, pasture, Christmas trees (WA, OR)
	Roads and reservoirs (CA, OR, WA, UT, ID)
	Urbanization (CA, WA, OR, CO)
	Second homes (CA, WA, OR, CO)
	Reservation for parks or wilderness (CA, WA)
	Reclassification to noncommercial forest (OR, CA, WA)

Source: U.S. Forest Service, forest surveys taken during the period 1965-81, and interviews with regional forest officials. Because many of the surveys did not have exact statistics on the amount of land converted to each use, we have not attempted to rank these causes in order of importance.

In the South, agriculture is the major competitor for forest land. In the last decade or so, agriculture has claimed 1.1 million acres of commercial forest in Arkansas, 1.4 million in Louisiana, 1.3 million in Mississippi, and 450,000 in Florida (Van Hees 1980, Murphy 1975, Murphy 1978, Bechtold and Knight 1980). Some of this land lies in the floodplains of major southern rivers. Formerly occupied by bottom-land hardwood forest, the land has proved highly suitable for the production of soybeans, a crop that can tolerate moderately wet soils. After the end of World War II, world demand for soybeans for use as animal feed or as edible oils soared. But the supply of suitable land in the Corn Belt was limited, particularly since the demand for corn has also been strong. Southern farmers found that they could buy bottom land for a few hundred dollars per acre, invest substantial sums in land clearance and drainage, and still obtain fertile land for a fraction of Corn Belt prices (Bickers 1978).

Elsewhere in the South, forest land is being cleared for pasture. Until recently, expansion of the southern livestock industry was hindered by chronic problems with disease and parasites. New drugs have removed this limitation, and a livestock expansion now under way is projected to result in a 70 percent increase in southern grazing by the end of the century (U.S. Forest Service 1980). Indirect impacts of urbanization appear to be playing only a small role in this process: during the 1967-75 period, 770,000 acres of southern pasture were urbanized or inundated, while 8.4 million acres of forest were converted to pasture.

Urbanization, roads, and reservoirs have been second only to agriculture in claiming southern forest land. Population in the southern region grew by 19 percent in the last decade, close to twice the national average. The sprawling suburbs of many expanding southern cities have moved into contiguous forest land, while a widespread regional preference for wooded second-home and hunting and fishing parcels has increased settlement of forested rural areas. Construction of the interstate highway system has also affected forest land in the South. One southern timber company (Union Camp Corporation), for example, found so much of its Georgia forest lands affected by the construction of Interstate 95 that it set up a subsidiary to develop properties surrounding interchanges.

Finally, close to a million acres of southern forest land have been turned to water uses. Some conversion has been the result of very large reservoir projects, such as those along the Savannah River in South Carolina and Georgia and along the Sabine River in Texas. Other forest land has been lost to a host of small impoundments created for recreation, fishing, or water storage. Although few large new dams are now being built, two long-planned exceptions, the Richard B. Russell Dam on the Savannah River and the Tennessee-Tombigbee Waterway in Alabama, would affect 53,000 and 56,000 acres of forest land respectively.

In a handful of southern states (Virginia, South Carolina, West Virginia, Tennessee, Kentucky) abandonment of marginal cropland has continued at a rate high enough to offset the loss of forest land, resulting in a stable or increasing total forest area. Except in Kentucky and West Virginia, these offsets are not expected to continue (Wall 1982).

In the North, Forest Service survey data tend to support NALS/SCS figures showing that urbanization is the greatest cause of forest land loss in the Northeast, while agricultural conversions are most important in the Midwest.[7] For example, a Pennsylvania study found that of the 600,000 acres going out of commercial forest in the 1965-78 period, nearly 60 percent became classified as 'urban and other' (Powell and Considine 1982). In Maryland, twice as much forest

7. NALS/SCS reported that in the Northeast, urban and water uses consumed 1.4 million acres of forest land, while crops and pasture consumed 2.7 million. But reversion of crop and pasture land to forest almost completely offset the agricultural conversion. In the Lake States and Corn Belt, however, crops and pasture removed 3.9 million acres (net), urban and water uses only 694,000.

land was cleared (1961-72) for housing as for agriculture (Bones 1980). In Minnesota, by contrast, agricultural uses took about the same amount of land (1962-77) as did urban uses (Jakes 1980).

Although the northeastern and midwestern states experienced relatively slow population growth during the 1970s, virtually all of their net growth took place in nonmetropolitan areas. This sort of growth has a disproportionate impact on forest land. First, many of the region's fastest growing nonmetropolitan areas are forested—for example, rural parts of Vermont, New Hampshire, and Michigan's Upper Peninsula. Indeed, the amenities of a rural, forested setting have been very important in attracting urban-rural migrants in the northern states. Second, nonmetropolitan growth frequently involves settlement at lower densities than are typical of cities, or even suburbs. In the northern states, a few rural settlers seek economic independence through subsistence farming; many more pursue farming as a hobby while working at off-farm jobs. Even for people with no interest in cultivating land or engaging in forestry, low rural land prices have encouraged residential settlement of land parcels of 5 to 40 acres. Zoning and subdivision control laws in a number of northern states have further encouraged this practice. For example, in Michigan, a proliferation of 10.1-acre lots has resulted from a state subdivision law that has a 10-acre cutoff and local zoning that commonly imposes a 10-acre minimum. According to one observer in that state, 'What we have accomplished by this is thousands of lots where a person uses a half-acre for a house, while 9½ acres lie idle' (Healy and Short 1981).

Second homes in forested settings are popular in the northern states, as they are in the South. The height of the second home boom in most areas occurred during 1969-73. Even timber companies, including Boise Cascade, Scott Paper, and International Paper, plunged into the recreational development business. Some of these developments were on areas formerly used as timberland. The 1973-75 recession bankrupted many rural developments, and many of the timber companies greatly reduced their operations. In subsequent years, recreational interest in rural land has focused less on the large-scale 'developments' than on individual purchases of rural acreage. Rural real estate dealers frequently cater to this demand by splitting up larger parcels into more easily marketed sizes.

In Ohio, Vermont, and Illinois, farmland reversion has more than offset urbanization, and the forest land base has increased. Only in Ohio is this offset expected to continue (Wall 1982).

In the West, one of the largest competitors for timberland has been wilderness and parks. (This use has also increased in such states as Georgia, Minnesota, Michigan, and Missouri.) Wilderness designation maintains or enhances the land's usefulness for ecological protection and for forest recreation, but bars the harvesting of timber products. Most of the wilderness reservations have been on federally owned lands, although some land has gone into state parks or nature preserves owned by nonprofit groups. The campaign to create wilderness is not necessarily centered in urban areas. Indeed, one survey of remote-area camping use found rural residents disproportionately represented (U.S. Department of the Interior 1973). But surely there must be some relation between the geographic spread of our urban civilization and the widespread perception of the need for a wilderness counterpoint.

Available data for the western states tend not to distinguish between agricultural and urban conversion of forest land. Both are reported to be occurring, although apparently not to the extent observed in the eastern part of the country. In California, for example, total forest conversions for residences, grazing, roads, reservoirs, and powerlines over a 22-year period (1953-75) were 202,000 acres or only 1.2 percent of the 1953 forest land base (Bolsinger 1980). In fast-growing western Oregon, the rate of conversion (1961/62-1975/76) was one quarter of one percent per year, with the bulk of the land going into pasture (Gedney 1981). About 13 percent of the land converted went to urban and suburban homesites. In Washington State, the spread of suburbs and second homes has affected a considerable amount of forest land in the Puget Sound and Olympic Peninsula areas (Bolsinger 1971, 1969). However, widespread ownership of land by the government and the forest products industry may protect some of the most highly productive areas.

In the Rocky Mountain West, there has not been much land clearing for agriculture because the timberland is generally not physically suited for agricultural use. The major source of conversion has been second homes (Van Hooser 1982). Large tracts have been subdivided into much smaller ones, and although many trees are left on the land, the affected areas are taking on the appearance of suburbia. Urban expansion is also occurring around metropolitan areas, with many states in the region recording population growth rates far higher than the national average. As in the Pacific Northwest, urban expansion is often contained when it encounters publicly owned forest lands.

Water impoundments have also been a major reason for the loss of forest land in the Rocky Mountain states, principally for large publicly constructed reservoirs (Van Hooser 1982). In general, however, the lack of agricultural clearing has meant that forest land conversion in the Rocky Mountain West (other than for wilderness) has been less than elsewhere in the country.

In evaluating the overall impacts of our urban civilization on forest land conversions, one might ask whether past urbanization pressures will continue. It is likely that some of the urban land uses prominent in forest land loss in the recent past will become less important in the years ahead. The interstate highway system is now virtually complete. And a combination of environmental pressure, lack of good dam sites, and high interest rates means that large-scale water projects will not be numerous. (On the other hand, increased use of irrigation may cause more small on-farm water impoundments to be created.)

Future conversions of forest land also depend on future land use controls. If states and localities restrict the urbanization of prime agricultural land, some of that growth could go to forested tracts. But this mechanism will most likely be more than offset by the fact that smaller loss of cropland means lower 'replacement' demand for agricultural use of forest land.

Most important, yet most uncertain, is the outlook for continued rural settlement. The continued growth of the retirement-age population, high suburban housing costs, and the increasingly diverse mix of rural jobs indicate that the settlement trends of the last decade may continue unabated. This could put new concentrations of people in and near U.S. forests. Calvin Beale, chief demographer for the U.S. Department of Agriculture and authority on the rural population revival, believes that rural population growth has occurred disproportionately in forested areas of the country. He notes: 'All one has to do is to compare a map of rural population growth with a map of land cover and the coincidence of growth and forests stands out. "Back to the landers" are not moving to the Great Plains nor the heart of the Corn Belt; they are settling in places like the Upper Peninsula (of Michigan), or Vermont, or the Blue Ridge or the Ozarks' (pers. comm. August 1982).

As I have already explained, impacts of urbanization must be considered in the context of all the conflicting demands on the forest land base. A recent U.S. Forest Service study has attempted to project what might happen to the total stock of commercial timberland over the next fifty years (Wall 1982). Based on consensus estimates by many regional Forest Service researchers, the study tries to project the net impacts of urbanization, agricultural conversions and reversions, wilderness reservations, and other factors. According to the projection, the nation's commercial timberland base could fall from 482 million acres in 1977 to 446 million in 2030, with more than half the drop coming in the next twenty years. The greatest fall (11 percent) is expected in the Pacific Coast states (excluding Alaska), while the smallest (5 percent) is expected in the North. Much of the decline in the Pacific Coast area appears to be due to expected wilderness designations, so that the land will remain in forest even though statutorily removed from timber management.

These numbers, if they can be relied upon, are comforting. But a complete analysis of urbanization and the land base must ask additional questions. Perhaps the most important one concerns land quality. If changes in the land base are the net result of conversions of forest to other uses and the reversion of nonforest land to forest, might not the average quality of woodland be getting better or worse over time?

This issue could become particularly important in light of the apparent slowdown in the amount of farmland abandoned and allowed to revert to forest. In past decades, farmland reversion was probably an important force raising the average quality of the timberland base. Land marginal for farming often makes relatively good timberland; moreover, in many areas of the country, reverting farmland regenerates in softwoods, making farmland reversion an important source of the trees now being cut in such places as New England and the Southeast. In addition, several million acres of reverted farmland in the South were planted in pines under various federal programs, and are also important in the region's current harvest. But there are indications that this replenishment of forest quality has come to an end.

Between 1954 and 1964, total U.S. cropland fell by 21.5 million acres. About half of this land was in the South and much of it grew up into pine forest. But then cropland began to increase again and by 1978 stood at nearly 27 million acres over the 1964 level. Again, the southern states accounted for about half of the gain. A Forest Service researcher in that region describes the implications for forest land quality: 'Better, flatter, higher quality forest land is always being looked at for other uses—for housing, for pasture, for crops, for fruit growing. Forests are always being pushed back into the mountains and the swamps. Meanwhile, abandonment of farmland has almost stopped, and what is reverting to forest now is coming from very miscellaneous sources' (McClure 1982).

The most obvious way to measure the net effect on forest land quality is to look at changes over time in the average 'site class,' a measure of the number of cubic feet of wood per acre that land is capable of growing when in fully stocked natural stands. I therefore compared the distribution of site classes of commercial forest land, as reported by the Forest Service for 1970 and 1977. Table 4 shows what I found.

These figures appear to show a startling decline in the quality of the U.S. commercial timberland base. They indicate that nearly all of the net decline in timberland was caused by loss of land in the two highest site classes, the forest equivalent of 'prime' farmland. Over a period of just seven years, the stock of highest quality timberland (site class 85 +) fell by over 13 percent.

Examination of another source of land quality data, however, appears to contradict this finding. The Forest Service obtained the 1970 and 1977 figures not by national surveys but by aggregating data for states and substate regions obtained in periodic forest surveys. These surveys are taken on a staggered basis, with each area covered about once a decade or so. The 1970 and 1977 data were actually adjustments to a common base year of the most recently available forest survey for each area.

In order to eliminate any adjustment error, and to obtain more recent figures, I analyzed ten state surveys taken since 1977 and compared each with the latest previous survey of the same area. (The areas were South Carolina, 1968-70; southwest Georgia, 1971-81; Arkansas, 1965-78; Florida, 1970-80; southeast Georgia, 1971-72; Pennsylvania, 1965-78; Ohio, 1968-79; eastern Oregon, 1963-78; Puget Sound, Washington, 1967-80; and Olympic Peninsula, Washing-

Table 4. Commercial Forest Land Area by Site Class (thousands of acres).

	U.S. Total	120+	85-120	50-85	20-50
1970	489,002	49,867	113,550	194,282	131,299
1977	471,336	45,828	96,273	197,327	131,907
Change	-17,666	-4,039	-17,277	+3,045	+ 608
% change	- 3.6%	- 8.1%	- 15.2%	+ 1.6%	+0.5%

Sources: U.S. Forest Service 1974, 1980. Data for both years exclude 5.0 million acres of national forest land in the Rocky Mountain states that were not included in the base allowable cut in 1970 because of such factors as unstable soils, small size of isolated patches and stringers, or special use constraints, and were subsequently removed from the commercial forest land classification.

ton, 1966-80.) I found that average site class *rose* in nine cases out of ten. Regional bias does not appear to explain the difference: site class rose in four out of five southern areas, two out of two northern areas, and three out of three western areas.

Conversations with persons responsible for regional forest surveys indicated that relatively large areas of land were being reclassified from one site class to another between surveys as a result of changing estimates of their potential productivity. For example, tens of thousands of acres of low quality land in south Florida have been moved in and out of the commercial forest land category as a result of changing ideas as to their timber potential. This kind of reclassification obscures any changes in land quality that may be due to forest clearing and reversion. Given the ever increasing importance of land quality to intensive forestry, one can only hope that the Forest Service will give more attention to obtaining consistent intertemporal data on land quality.

OTHER URBAN IMPACTS ON FOREST LAND

Outright conversion is not the only way in which an urban civilization affects forest land. There are also several other, somewhat more subtle, impacts. Most, but not all, stem from urbanites' increased interest in owning, and sometimes building on, rural land. In work published elsewhere (Healy and Short 1981) I have documented three trends that have, over the last three decades or so, characterized the rural land market nationwide. Although they are characteristic of all types of rural land, each appears particularly pronounced in the case of forest land. The trends are (1) increased ownership of forest land by so-called nontraditional owners, including urban investors and new, nonfarm residents, (2) higher forest land prices, and (3) division of forest land into smaller parcels. Other recent literature (McCarthy and Morrison 1979) provides abundant material on a fourth trend—the recent reversal of long-standing demographic trends that has brought millions of new residents to small towns and rural areas.

In analyzing the impacts of these economic and demographic trends, it is convenient to divide them into three groups: impacts associated with new owners and higher land prices, impacts associated with parcel size, and impacts of the physical juxtaposition of people and forest land.

New Owners and High Prices

The most significant change in forest landownership since the end of World War II has been the decline in the amount of wooded land owned by farmers. In 1952, farmers owned 172 million acres of forest land; by 1977, their holdings had fallen to 116 million acres. Some of this land has been converted to nonforest, nonagricultural use. But much of it has been transferred to 'miscellaneous' nonindustrial private owners. This 'miscellaneous' group now represents the largest single category of forest landowners. Some members of this group are urbanites who have inherited land from rural relatives; others have bought forested parcels for recreation or future retirement; still others have bought the land as an investment.

Nonindustrial private forests (NIPFs) include a considerable amount of potentially productive land; for example, they account for 71 percent of the timberland in the South. (This figure includes both farm and nonfarm NIPF holdings. Most available data make it impossible to distinguish the two.) Because of their aggregate size and productive potential, the NIPFs will have to play an extremely important role if future timber demands are to be met. They also provide a significant share of total forest recreation, particularly in the eastern half of the United States.

Foresters have long bemoaned the alleged low productivity of the NIPFs, pointing out that they are growing wood at much less than biological potential and have not been fully participating in the movement toward more intensive management that has become commonplace in the forest industry. Dozens of reports, task forces, symposia, and congressional hearings have considered the NIPF 'problem' (Sedjo and Ostermeier 1978, U.S. Forest Service and NASF 1979).

The reasons most frequently cited for the relatively poor performance of the NIPFs are (1) insufficient financial return to owners' investments in timber production, (2) owners' ignorance of opportunities in forestry, and (3) owners' preference for using land to satisfy objectives other than timber production. The last two factors are directly affected by the changing composition of landowners.

It is difficult to fault landowners for not investing time and money in wood production when doing so would be unprofitable. But rises in the prices of forest products, combined with possible profits from firewood production and new tax incentives, appear to have improved the economics of timber management. Given the apparent profitability of increased wood production, are NIPF owners prepared to respond? Economist Marion Clawson thinks so. He argues that, over the last twenty-five years, nonindustrial private forests have made substantial gains toward matching the forest products industry's level of annual wood growth per acre. He attributes the NIPF's improved performance at least in part to the incentive of rising timber prices. Clawson argues bluntly that 'for the nonindustrial private forests, the best public action is to leave them alone, aside from fire control' (Clawson 1981).

Others who have studied NIPFs put more emphasis on the role of owner ignorance and of the owner's nontimber objectives in reducing the investment response to rising prices. For example, a study by the forest industry notes as a 'major restraint' to improved wood productivity the 'lack of knowledge by small private owners concerning the benefits of forest investment' (Forest Industries Council 1980, p. 45). Even Clawson concedes that economic models of timber output's response to price suffer 'the very serious weakness that forest owners/managers may not respond to price or other stimuli in ways or to the degree that the analyst thinks would be rational.' The changing patterns of forest landownership could thus become an important influence on investment in intensified forest management, hence on future wood output.

Although there have been many surveys demonstrating that personal characteristics of owners affect their attitudes to forest management and willingness to invest in it (Kurtz and Lapping

1976), there have been very few studies relating both economic return and owner characteristics to the investment response. One study that did so for a sample of New Hampshire landowners found that the probability of timber harvest was strongly responsive to stumpage prices, but farmers were more than twice as responsive to price than nonfarmers were (Binkley 1979).

Factors influencing timber harvest are not necessarily the same as those influencing investment in growing trees. One might speculate that to the extent that nontraditional owners have higher incomes, hence more to invest, as well as more education than traditional owners, they might be more receptive to taking advantage of profitable opportunities in tree planting or timber-stand improvement. Moreover, the new owners' feelings about the environment might predispose them to management options that would raise wood growth or improve the quality of such nonwood outputs as wildlife and water. Yet the idealistic desire to improve their land can falter when nontraditional owners face the hard work and expense that land management requires. One West Virginia forester I spoke with claimed, 'You can get [the newcomer's] ear a lot better than the old timer's.' But a colleague added, 'If it requires their own labor, you may as well forget about it.'

Foresters who deal with the rural public believe that landowners who have purchased forest tracts for recreation or investment are frequently misinformed about both the economics of timber management and how it might fit into their plans. Many new, urban-oriented owners tend to view resource protection in preservationist terms rather than in terms of 'conservation for use.' This leads the new owners to 'conserve' the forest by not harvesting any of the trees on their property. Yet, in many parts of the country, foresters point out that the wood being conserved is of very low quality, the result of decades of 'high grading' (selective cutting of the best timber), and it will not be much improved by the passage of time. They believe that prudent long-term management would involve heavy immediate harvesting and regeneration in sounder trees and more salable species.

In part, the strong preservationist inclination of many new owners may reflect an ignorance of forestry; it may also, however, reflect basic motives for buying rural land. The trees that shade a wooded recreational site are likely to have greater value to the owner as scenery than they would to a logging contractor as sawtimber.

The impact of preservation-minded landowners on wood supply is also affected by the likelihood that intentions with regard to timber harvest may change (Turner et al. 1977). A 1970 survey of Delaware woodland owners found that 58 percent said they had no plans ever to harvest timber. A 1974 resurvey found that although the proportion of owners who said they would never harvest remained constant, 17 percent of the original respondents had changed from opposition to harvesting to a willingness to harvest. Moreover, some parcels had been transferred by sale or inheritance during the four years between surveys, increasing the possibility that any given parcel would at some point in its history come into the hands of an owner interested in harvesting timber. The authors of the study also noted that landowners unlikely to harvest tended to own smaller than average tracts, reducing the potential impact on wood supply. Despite the probability of eventual harvest, however, the productivity of at least some land is reduced during whatever time it is withdrawn from active timber management.

When they do decide to harvest, nontraditional landowners may contribute to poor forest management in yet another way. They may allow irresponsible timber operators to 'high grade' their land or to harvest in a careless and destructive manner. A West Virginia forestry professor points to two counties noted for new urban buyers as 'good examples of people buying up land that they do not relate to in a stewardship sense. Much of our timber is nearing harvestable age and timber operators are approaching the landowners. We see a lot of harvesting done in a terrible manner. It makes a [substantial] and lasting water-quality problem; the land may be so eroded that it will take a long time to restore any cover; and it removes all economic value of the land [for forestry]' (Healy and Short 1981). Inexperienced urban owners who try to harvest firewood, an important factor in many hardwood areas, may also degrade the forest through careless cutting.

Relatively few of the impacts of nontraditional ownership are likely to be felt in the near term. Little of the land purchased for recreational use (with the exception of some softwood areas of the South and Pacific Northwest) is currently a major source of wood fiber supply. But lack of immediate need is no reason for complacency: it takes up to thirty years to produce good softwood sawtimber, and up to a hundred for quality hardwoods. To the extent that owner ignorance or inattention causes underinvestment in nonindustrial private forest lands, future generations may face unnecessary shortages of wood products.

Increased demand for forest land by nontraditional owners has been one (though not the sole) cause of the dramatic increases in land prices observed over the last three decades. Price data are incomplete, and tend to be anecdotal. However, one set of national data obtained from the Forest Service revealed that between 1969 and 1979 the average per acre price of timberland purchased by that agency with Weeks Act funds rose by more than 500 percent, while the price of recreational forest land bought with Land and Water Conservation Fund monies rose more than 700 percent (Healy and Short 1981).

As an economist, I am not inclined to term the high relative price of anything as a 'problem.' Rather, it is simply an indication that a commodity is scarce and sought after. However, high land prices do appear to be a problem from the standpoint of two important parts of the forest constituency: (1) public agencies charged with providing outdoor recreation, who find that high land prices severely strain limited budgets, and (2) timber companies that seek to add to their land base, yet are unprepared to compete with recreational and residential buyers. Fortunately for the companies, there seems to be a continued sharp divergence between the per acre price of medium-size tracts and that of extremely large (say 1,000+ acres) tracts, with the large parcels selling at far lower per acre prices. As long as large tracts are available, companies will be able to expand their holdings at prices that bear a reasonable relationship to the returns from timber growing. Changing parcel sizes, however, may in the long run start to erode that opportunity and lead the companies into a difficult head-to-head competition with nontraditional owners.

Parcellation

Accompanying the acquisition of forest land by nontraditional owners has been the division of many tracts into parcels of smaller size, a process known as 'parcellation.' The size of parcel has traditionally been determined by custom (for example, the 40-, 80-, 160-, and 640-acre plots resulting from 18th and 19th century government land surveys) and by the economics of past land uses. As economic forces change, the size of parcels slowly changes to reflect them.

In many areas of crop agriculture, parcel sizes have been increasing as a result of scale economies in modern agriculture. But in many timbered areas, particularly where the number of nontraditional owners is growing, parcel sizes appear to be decreasing. Recreationists, rural nonfarm residents, and land investors seem most interested in parcels in the 5- to 40-acre range, and local landowners and real estate agents have made a profitable business of splitting parcels to meet the new demand.

Data on the rate of parcellation are scarce. But in research conducted recently at the Conservation Foundation, I studied six representative rural counties, scattered across the United States

(Healy and Short 1981). Three of these could be described as predominantly forested: Hardy County, West Virginia; Plainfield, New Hampshire; and Tyler County, Texas. The three forested counties exhibited the highest rate of parcellation in the six counties studied, with annual compound rates of growth in number of parcels over the period 1954-76 ranging from 6.1 percent (in Tyler County) to 2.4 percent (in Hardy County).

It is important to emphasize that the relatively rapid creation of small parcels may not quickly cut into the total land base. Indeed, in each of the three forested counties, the bulk of the land remained in relatively large tracts. But the direction of change, as well as the problems already evident in some local areas, indicates that parcellation should be viewed as a potentially serious threat.

Recent data allow us to characterize the extent of the parcel-size problem for forest land nationwide. Data for 1977 compiled by Linda Lee of Oklahoma State University revealed that 22 percent of all privately owned U.S. forest land is held in ownership units of less than 100 acres; 11 percent is held in tracts of less than 50 acres.

Foresters disagree on what constitutes the minimum parcel size for efficient timber management, but it is clear that the cost of harvesting timber is greater on small tracts. A timber buyer for a large lumber company in Texas observes: 'We have people coming to us all the time trying to sell timber in 2-acre, 3-acre, and 5-acre blocks. We just refer them someplace else. We hardly ever buy anything less than 25 acres, and we prefer 50 or more. Our contractor loses a day's work just moving his equipment to the next site.'

Even more serious is parcellation's impact on efforts to improve productivity in management methods. Forest economist Clark Row calculates that tract size has a substantial impact on the financial returns to intensive forest management, and that scale economies are found as parcel size rises, up to at least 160 acres. He concludes that 'where tracts are small, owners may accurately perceive that *for them* intensive timber growing would not be worth the effort, unless it facilitates other objectives' (Row 1978).

Moreover, even if a given type of management shows a high percentage rate of return on investment, the revenue produced on a small tract may not be large enough to make a landowner take action. For example, few urban absentee owners would be motivated by a $100 per year incremental return from thinning a hardwood timber stand, even if the rate of return on investment was quite high. The same thing applies to nontimber uses such as leasing hunting rights or other kinds of access rights.

A recent study of forest land in Oklahoma demonstrated that parcel size had a measurable influence on the intensity of forest management (Thompson and Jones 1981). The researchers found that styles of management tended to differ depending on whether parcel size was in the range of 10 to 50 acres, 51 to 700 acres, or 700-plus acres. The acreage divisions between the groups were not arbitrarily fixed, but were set at the level at which the type of management began to change. As might be expected, the larger the tract size, the more likely the owner was to be interested in wood production, the intensity of management was higher, and the probability of harvest was greater.

Problems of Juxtaposition

I suspect, though I have no way of proving it, that more people live and play in and around America's forests than at any time in our history. Two rough indicators are that in 1980 rural population set a new record (59.5 million) and so did recreational use of the national forests (233 million visitor days). And the presence of people seems to be growing in many forested areas of commercial importance. In one hundred of the country's leading timber-producing counties, population growth (1970-80) was more than twice the national average rate.[3] This increasing juxtaposition of people and forests is creating problems for forest managers and for the people themselves.

Perhaps the most obvious problem is the aesthetic change that dispersed urban uses—sometimes called 'buckshot urbanization' (Little and Fletcher 1981)—are bringing to the rural forested landscape. The growth of the rural nonfarm population has led to a proliferation of houses along rural roadsides. They are often accompanied by schools, factories, and commercial strips. It does not take much of this sort of development to make the countryside begin to appear 'filled up,' even when the actual density is quite low. Even many public lands are suffering aesthetic damage from nearby urban uses, particularly through the development of privately owned 'in-holdings' or the building of resort communities on the periphery of public recreation areas (Shands 1979).

For recreationists, juxtaposed urban uses mean that it is increasingly difficult to secure access to private lands for hunting, fishing, and hiking. Says the chief warden of Maine's state wildlife department: 'There are more and more people coming into the state all the time and a lot of them are buying up property. Often, land that has been open for years is suddenly posted by its new, out-of-state owner' (Portland, Maine *Press-Herald*, August 30, 1977).

The building of second homes along its length has forced the rerouting of several segments of the Appalachian Trail. For the timber industry, it can be more difficult to maximize timber yields when there are nearby residents or nontraditional property owners. In a number of cases, forest residents have objected to aerial spraying for insect or weed control. Elsewhere, timber managers planning clearcuts near rural residences have met with threats of lawsuits or vandalism.

Increased urban development in and around forest areas, as well as greater recreational use of forest land, has also created problems concerning fire suppression and prevention. Because over 90 percent of woodland fires are caused by humans, increased numbers of people in the woods simply offer more opportunities for such classic fire initiators as careless smoking, children playing with matches, and sparks from automobile and motorcycle mufflers. Vacation homes, vacant for part of the year and often built under laxly enforced or nonexistent codes, afford opportunities for electrical fires that can spread to the woods. Scattered dwellings also mean long response times by firefighters, so that once fires start they may have considerable time to spread.

Arbor Vitae, Wisconsin, for example, is an isolated rural community whose population has doubled since 1970. Its many summer residents spend only a few weeks a year there, and a growing number of retirees have built homes in the woods or around lakes. The homes are usually of wood frame construction and, for charm, often have shake shingles. The area is frequently snowbound in winter, making access difficult. The fire chief describes the community as 'a fire protection nightmare' (Harrison 1979).

Greater numbers of people living in woodland areas have also put new demands on fire departments to protect life and structures, in addition to the woods themselves. This means that the traditional 'perimeter strategy' of fighting forest fires by containment often must be sacrificed in favor of protecting scattered structures. Says one forest fire authority, 'This puts forest fire

3. The counties were the one hundred highest in the nation in value of forest products sold from farms in 1974. (Data for total forest products sales by counties were not available, but many of the counties on the list also have very large government and industry sales.) The population of the one hundred counties grew by 24.2 percent during the last decade, against a total U.S. population growth of 10.8 percent.

agencies and fire departments at a disadvantage, since most training in the past has concentrated on perimeter strategy. Additional [problems] arise as the state, city, and county fire departments confront the extraordinarily flashy grass, brush, and timber fuels in the urban-wildland border' (Wilson 1977).

In addition to making firefighting more difficult, settlement of forested areas has raised the question of who should take responsibility for fire protection. In the Sierra Foothill region of California, a wooded area that has been growing very rapidly in the last several years, the state planning agency reports, 'the expanding use and development of the wildlands are causing difficulty in providing adequate fire protection. In particular, unclear local, state, and federal control of new development makes fire protection unmanageable' (California Office of Planning and Research 1980, p. 1). The main problem, in California and elsewhere, is that local governments have depended on state and federal forestry agencies to fight all fires in wooded areas. But these agencies are designed to fight forest fires, not structure fires. They face a serious dilemma: they cannot ignore structure fires, if only because they cannot allow them to spread to adjoining woodlands. Yet if they do respond to such emergencies, local authorities will increasingly rely on them for general rural fire protection. Says the California report, 'The [California Department of Forestry's] willingness to respond to structural fires may be encouraging local governments to relinquish their responsibility for providing fire protection to the residential developments they continue to permit' (p. 4).

Air Pollution

Urban areas, as well as rural power plants and industrial concentrations, generate a variety of air pollutants. A number of instances of severe forest damage have been documented in the vicinity of power plants, copper, aluminum, and nickel smelters, and even a pulp mill (Smith 1981). Less dramatic but still considerable has been the impact of photochemical smog from the Los Angeles area on the mortality of conifers downwind in the San Bernardino National Forest.

Far more difficult to document, but much more worrisome, are the possible impacts on forests of lower levels of pollutants transported over long distances from the point of generation—often a major metropolitan area or power plant complex—to a wholly unindustrialized forest area. Smith (1981) concludes that over the next decade air pollution 'has the potential to cause significant reductions in forest productivity and shifts in species composition in numerous areas throughout the temperate zone' (p. 357). He calls air pollution, 'one of the most significant contemporary anthropogenic stresses imposed on temperate forest ecosystems' (p. 360).

Among the major problems are long-range sulfur dioxide transport from coal-fired power plants, ozone pollution downwind from major metropolitan areas, and particles of sulfate and trace metals from coal burning and from diesel engines. Perhaps the most frightening aspect of the effect of air pollution on forests is how little understanding there is of the transport of pollutants and the specific interactions with plant growth and mortality. For example, individual trees respond in very different ways to a particular pollutant: some are severely damaged, others are unaffected.

REFLECTIONS ON THE ROLE OF LAND IN FORESTRY

Gifford Pinchot defined forestry as 'to grow trees as a crop.' Yet, when compared with crop agriculture, current U.S. forestry practices are primitive. Site preparation, thinning, and planting—staples of crop agriculture since Neolithic times—have become common on industry lands only in recent years, and are still the exception on many nonindustrial private tracts. Fertilization and aggressive pest control—standard practices of modern agriculture—are uncommon in forest management. Genetic improvement in forestry is still accomplished mainly through progeny selection rather than hybridization, and in any case the supply of improved seedlings is limited (Hyde 1980). There is little doubt that in terms of *biological potential* the adoption of a more agricultural approach to forest management could work improvements in per acre yields equal to or exceeding those achieved in the recent experience of American crop agriculture. Can forestry follow the lead of crop agriculture and sidestep its land constraint through intensified management and increased productivity?

As usual, the answer depends in great part on economic costs and returns. It appears that, at recent timber prices, a great deal of intensified forest management is profitable. The Forest Industries Council, for example, has identified 139 million acres on which increased silvicultural investment would yield a 10 percent or better after-tax return (Forest Industries Council 1980). The average incremental yield over the twenty-five states studied was an impressive 78 cubic feet per acre per year. Since the current annual growth in all timberland in those states averaged only 51 cubic feet per acre, the startling implication is that for each acre on which the average economically justified investment was made, more than 1.5 acres of average productive land could be 'retired' from the forest land base. Actually, the prospects at the margin for maintaining production while retiring land from the timberland base are somewhat better than the 'average' example, because one could initially concentrate investments on land with the highest incremental yield, while retiring land with below-average growth.

The intensification of forest management has implications for the type of land that will be used for growing trees and for its location. Throughout most of the history of U.S. forestry, it was customary to emphasize trees, not land. The value of a given piece of woodland was in most cases set by the value of its existing timber stock, with little reference to the soil's inherent capacity to produce. But in recent years, timber companies and private investors have begun to appreciate that more fertile sites not only produce faster-growing natural stands but give higher returns per dollar invested in such silvicultural practices as planting, site preparation, genetic improvement, pest control, and thinning. (Fertilization, which disproportionately benefits some lower quality sites, is an exception.) William Hyde (1980) has documented how the most economically rational way to increase timber supply in the Douglas-fir region would be more intensive management of the better quality sites, with little or no investment on sites of lower quality. A California study found that over 80 percent of the potential increase in yield from planting understocked land or replacing hardwoods with conifers could be realized by treating only 46 percent of the land in those conditions (Bolsinger 1980). And Marion Clawson has argued that more intensive management of high-site land in national forests could free nearly 30 million additional acres nationally for wilderness or other nontimber use (Clawson 1976).

The most profitable management opportunities are most frequently found in the South and in coastal portions of the Pacific states (Forest Industries Council 1980). Their existence is confirmed by observation of recent patterns of forestry industry investment: capital continues to flow to the more promising regions and the more fertile sites.

Perhaps the most worrisome impact of an urban civilization on its forests is the extent to which it might interfere with this process of management intensification and geographic concentration. If suburban growth or agricultural 'replacement' clearing compete disproportionately for high-site index forest land, the economic opportunities for forestry investment will be fewer than have been projected.[4] If agricultural clearing and rural settlement remain unusually high in the southern region, opportunities for regional forestry expansion will be curtailed. And if wood production cannot be intensified and concentrated, society will lose one of its best opportunities for preserving large tracts of low-site land for wilderness and wildlife.

Other impacts will occur whether or not society chooses to intensify production on its better forest land. If parcellation or buckshot urbanization interferes with scale economies, formerly profitable forestry opportunities will disappear. If continued air pollution affects tree growth, all of our yield projections will be incorrect and more land will be needed to produce the same amount of timber. If urban uses are scattered through the forest, a larger amount of land will be needed to provide the same amount of recreational value or of scenic amenities.

Whether these impacts occur is a choice—a series of choices, really—that must be made by all of us as members of an urban civilization. The forest constituency can point to potential injury, but the means of avoiding it lie well beyond the forest. The urban civilization can decide to put its homes and factories as much as possible on low-site land. It can protect its prime farmland and thereby reduce 'replacement' demand. It can resist the tendency toward residential scatteration. It can begin to address creatively the problems of nontraditional ownership and parcellation (Healy and Short 1981). Only then can it fully realize the great opportunities for the practice of a new, concentrated, and intensified brand of forestry, permanently maintained within the context of an urban civilization.

4. Although there are certainly instances in which forestry can compete with agriculture, they are uncommon, at least at current prices of crops and wood. Hyde estimates that the timber production value of good quality Douglas-fir land (yearly growth 120-225 + cubic feet/acre/year) at a stumpage price of $140 per unit is $500 to $1,500 per acre (Hyde 1980). That range of prices is low relative to those commonly paid for cropland or residential or recreational land in the region today.

REFERENCES

Bechtold, W. A., and H. A. Knight. 1980. Florida's forests. U.S. Forest Service Resource Bulletin. Southeastern For. Exp. Sta. Asheville, N.C.

Bickers, J. 1978. Why the Southern land boom may be just beginning. Progressive Farmer 93(7):15-17.

Binkley, C. S. 1979. Timber supply for nonindustrial private forests: An economic analysis of landowner behavior. Ph.D. dissertation, School of Forestry and Environmental Studies, Yale University, chap. 4.

Bolsinger, C. L. 1969. The timber resources of the Olympic Peninsula, Washington. Pacific Northwest For. and Range Exp. Sta., Portland, Oregon. U.S. Forest Service PNW-31.

———. 1971. Timber resources of the Puget Sound area. Pacific Northwest For. and Range Exp. Sta., Portland. U.S. Forest Service PNW-36.

———. 1980. California forests: Trends, problems, opportunities. Pacific Northwest For. and Range Exp. Sta., Portland. U.S. Forest Service, PNW-89.

Bones, J. T. Forest-land clearing and wood recovery in Maryland. Northeastern For. Exp. Sta., Broomall, Pennsylvania. U.S. Forest Service NE-290.

California Office of Planning and Research. 1980. Fires in the foothills. December (reproduced).

Clawson, M. 1976. The national forests: A great national asset is poorly managed and unproductive. Science 191 (4227):762-767.

———. 1981. Strategies for future forest production. Paper presented at Resources for the Future Conference on Forest Land Use, Washington, D.C., March 30.

Crosson, P., ed. 1982. Cropland crisis? Published for Resource for the Future. Johns Hopkins University Press, Baltimore.

Forest Industries Council. 1980. Forest productivity report. National Forest Products Association, Washington, D.C.

Gedney, D. R. 1981. Change in area and ownership of private timberland in western Oregon between 1961-62 and 1973-76. Pacific Northwest For. and Range Exp. Sta., Portland, Oregon. U.S. Forest Service PNW-92.

Harrison, A. 1979. Help for the rural fire chief—Part 1. Fire Command, March.
Hays, S. P. 1980. The role of forests in American history. Paper presented to the Weyerhaeuser Company Foundation Symposium on the Future of Forest Land Use, Seattle, Washington, April.
Healy, R. G., and J. L. Short. 1981. The market for rural land: Trends, issues, policies. The Conservation Foundation, Washington, D.C.
Herrington, B. E., and R. N. Shulstad. 1982. Conversion of delta woodland and pasture to cropland: Economic feasibility and implications. University of Arkansas, Agricultural Experiment Station. March.
Hyde, W. F. 1980. Timber supply, land allocation, and economic efficiency. Published for Resources for the Future, Johns Hopkins University Press, Baltimore.
Jakes, P. J. 1980. The Fourth Minnesota Forest Inventory: Area. North Central For. Exp. Sta., St. Paul, Minnesota. U.S. Forest Service NC-44.
Kristol, I. 1972. An urban civilization without cities? Current 145 (November):3-14.
Kurtz, W. B., and M. B. Lapping. 1976. The small forest landowner: Ownership and characteristics. Exchange Bibliography No. 1113. Council of Planning Librarians, Monticello, Illinois.
Little, C. E., and W. W. Fletcher. 1981. Buckshot urbanization: The land impacts of rural population growth. American Land Forum 2(4):10-17.
McCarthy, K. F., and P. A. Morrison. 1979. The changing demographic and economic structure of nonmetropolitan areas in the 1970s. Rand Paper P-6062. The Rand Corporation, Santa Monica.
McClure, J. 1982. Project Leader, Southeastern For. Exp. Sta., Asheville, N.C., personal communication, July.
Middlesex-Somerset-Mercer M Regional Study Council, Inc. 1980. Planning for agriculture in New Jersey. MSM Regional Council, Princeton, N.J.
Murphy, P. A. 1975. Louisiana forests: Status and outlook. Southern For. Exp. Sta., New Orleans. U.S. Forest Service SO-53.
———. 1978. Mississippi forests: Trends and outlook. Southern For. Exp. Sta., New Orleans. U.S. Forest Service SO-67.
National Agricultural Lands Study [NALS]. 1981. Final report. U.S. Government Printing Office, Washington, D.C.
People for Open Space. 1980. Endangered harvest: The future of Bay Area farmland. People for Open Space, San Francisco.
Powell, D. S., and T. J. Considine, Jr. 1982. An analysis of Pennsylvania's forest resources. Northeastern For. Exp. Sta., Broomall, Pennsylvania. U.S. Forest Service NE-69.
Raup, P. M. 1980. Competition for land and the future of American agriculture. In: S. S. Batie and R. G. Healy, eds., The future of American agriculture as a strategic resource, pp. 41-78. The Conservation Foundation, Washington, D.C.
Row, C. 1978. Economics of tract size in timber growing. Journal of Forestry 76(9):576-582.
Sampson, R. N. 1982. Farmland or wasteland: A time to choose. Rodale Press, Emmaus, Pennsylvania.
Sedjo, R. A., and D. M. Ostermeier. 1978. Policy alternatives for nonindustrial private forests. Society of American Foresters and Resources for the Future, Washington, D.C.
Shands, W. E. 1979. Federal resource lands and their neighbors. The Conservation Foundation, Washington, D.C.
Simon, J. L. 1982. Are we losing our farmland? Public Interest 67(Spring):49-62.
Smith, W. H. 1981. Air pollution and forests. Springer-Verlag, New York.
Thompson, R. P., and J. G. Jones. 1981. Classifying nonindustrial private forestland by tract size. Journal of Forestry 79(5):288-291.
Turner, B. J., J. C. Finley, and N. P. Kingsley. 1977. How reliable are woodland owners' intentions? Journal of Forestry 75(8):498-499.
U.S. Department of the Interior, Bureau of Outdoor Recreation. 1973. Outdoor recreation: A legacy for America. U.S. Government Printing Office, Washington, D.C.
U.S. Forest Service. 1974. The outlook for timber in the United States. U.S. Government Printing Office, Washington, D.C.
———. 1980. An assessment of the forest and range land situation in the United States. FS-345. U.S. Government Printing Office, Washington, D.C.
U.S. Forest Service and National Association of State Foresters. 1980. Proceedings of the National Nonindustrial Forestry Conference, November 26-27. General Technical Report WO-22. U.S. Government Printing Office, Washington, D.C.
U.S. Soil Conservation Service. 1977. Potential cropland study. GPO, Washington, D.C.
Van Hees, W. W. S. 1980. Arkansas forests: Trends and prospects. Southern For. Exp. Sta. New Orleans. U.S. Forest Service SO-77.
Van Hooser, D. D. 1982. Project Leader, U.S. Forest Service Inter-Mountain For. and Range Exp. Sta., Ogden, Utah, personal communication, August.
Wall, B. R. 1982. Trends in commercial timberland area in the United States by state and ownership, 1952-77, with projections to 2030. U.S. Forest Service WO-31. U.S. Government Printing Office, Washington, D.C.
Wilson, C. C. 1977. Fatal and near-fatal forest fires: The common denominators. International Fire Chief 43(9):10.
Woodruff, A. M., ed. 1980. The farm and the city: Rivals or allies? Prentice-Hall, Englewood Cliffs, N.J.

3

FOREST LAND USE AND SETTLEMENT: A GEOGRAPHER'S COMMENTS

Richard L. Morrill

Land, of course, is the basic subject matter of geography. At the risk of oversimplifying, the use and character of land are governed by two interacting processes—physical and social. In most of the world there is intense competition for the use of land, suggesting the general principle that whoever owns or controls land will tend to use it to yield the greatest utility over a reasonably long term. Physical processes largely determine the long-run productivity of the land. But the key theory of the human organization of space relevant to this symposium posits a gradual decline in intensity of use and land value with greater distance or decreasing accessibility to and from centers of control and exchange—that is, our large cities. As cities grow, they increasingly are able to transform and intensify the use of such accessible land. Forest land, like any other hinterland, is, from the standpoint of theory about the competition for space, inevitably subject to change. If society as a whole wishes to change the rules by constraining competition through planning and regulation, thereby preserving forestry, for example, then the demand for land for more intensive uses, and thus future production of value, is either lost altogether or transferred to land elsewhere, usually around a different city. The justification for such intervention will be discussed later.

The best technical definition of the urban/forest interface is perhaps that it is a situation in which the marginal value of the land is approximately equal for all uses—forestry, farming, or urban. Geographically, this tends to coincide with the outer portion of the urban commuter shed (around a city like Seattle, part of the zone from 15 to 30 miles out—that is, over a territory that may well be as large or larger than the urban area itself. However, pressures on forest land are not confined to the vicinity of cities, but are significant in some recreational areas and in mixed farming and forestry areas.

It is clear that land conversion—at least the notion that the loss of higher productivity forest land to urban, recreational, or wilderness purposes represents a serious erosion of the nation's resource base—is a nonissue. The opportunities for and expected success of intensification of growth on nonthreatened forests vastly outweigh the probable withdrawals, even including to agriculture and wilderness, of from 1 to 2 percent per decade. For example, in the Pacific Northwest, less than 1 percent of the land is in urban uses. Even if urban use were to double in the next several decades, this would not impinge on even 1 percent of the land in forests. Forestry, like fisheries, has scarcely advanced beyond 'hunting and gathering' technologies, as Robert Healy observes, since there has been no significant economic pressure or need to do so. One could as readily argue that, given the relatively low prices for wood products, we have too much commercial forest land. In other words, possible adverse impacts of conversion on supply and prices have not occurred. Certainly there are local areas or regions in which the impact of conversion is greater, not, I believe, because of urbanization, but mainly because of agriculture (in parts of the South) or recreation (in parts of California and Florida). As a geographer I concur strongly with Perry Hagenstein that the argument for intervention and subsidy to maintain regional self-sufficiency is far weaker than the argument for national intervention to preserve a national resource. Clearly there is an integrated national market for wood products, and efforts to protect regional self-sufficiency could result in serious and inefficient misallocation of resources.

Still, it is disturbing that the data are apparently so crude that it is not possible to be certain even about the direction of net change by quality class in specific regions, in part because of frequent definitional changes. There appears to be rather little of the highest productivity land under direct threat of urbanization, simply because there is usually sufficient lower quality (from a forestry point of view) land available for residential or recreational use. It would be useful to have some accurate inventories of just where and how much of the best forest land is converted, so that locally appropriate responses can be devised. As with conversion of farmland, crude aggregate national statistics are alarmist and misleading and do not lead to specifically useful policy. Also, data on rural residential or recreational settlement are anecdotal and incomplete. There is reason to hope that the 1990 census will for the first time permit the detailed mapping of rural settlement.

What of the concern or desire for forest as landscape and open space around our cities? Except for parts of the South, it should not surprise us that there is relatively more forest cover today than in the period of maximum clearance for agriculture, 1880 to 1920. We can hardly justify subsidy of forest ownership or exclusive forest zoning on grounds that the market fails to supply it. Thus, if society wishes to guarantee the continued existence of particularly aesthetic tracts, the only fair way is by outright acquisition, and this is quite routinely done in the development of parks.

The real issue then is whether there is a significant added cost and lower productivity and profitability of carrying on forestry in the vicinity of cities or prime recreational areas, and even if there is, whether this is sufficient reason for public intervention. Adverse impacts on forestry come from possibly higher taxes due to increasing demand for more profitable uses, greater management restrictions due to sensibilities and preferences of urban residents, environmental degradation from air pollution and fire, and diseconomies resulting from land fragmentation—for example, near city Department of Natural Resources school trust lands in Washington State. Competition of forestry with agriculture, recreation, and urbanization is hardly new. The real question is whether the magnitude and kinds of pressures are such as to justify greater land use controls on the urban-rural fringe or the subsidy of near-city forestry. There is no doubt that the impacts are real. There are temptations to subdivide near-city tracts when the return may be several times that gained from forest use. New owners *are* less likely to be knowledgeable about forestry management. The aesthetic preferences of nearby city residents can make it more difficult or expensive to apply preferred management practices. Risks from pollution and fire are increased. And certainly, where forest land is fragmented into numerous farms and gentlemen farms and parcels, productivity is quite likely to be reduced. And the impact is not going to lessen.

It is true, as Healy has emphasized, that rural settlement is very active, perhaps more than it has been for fifty or more years. While the greatest rate of growth in rural settlement is exurban—that is, in the outer commuter zone beyond the built-up suburbs—there has also been significant decentralization of employment and movement of people to small towns, villages, and scattered small subdivisions and individual residences across most of the country, except the Great Plains. My own studies of settlement preferences and migration indicate a tremendously strong American preference for low density living, a force that no amount of sermonizing about the efficiency of dense city living is going to change. And, contrary to conventional wisdom, very large cities are not more efficient than small cities, or towns, or rural settlement. At the margin, all the settlement sizes must be equally efficient, otherwise they would not continue to form and exist. And, contrary to what many think, the higher real costs of energy will encourage economic decentralization to suburbs of big cities and to smaller cities generally. Thus, more urban/forest interface pressures in the future are going to be felt around smaller cities than in the recent past, if there is a gradual reduction in long-distance commuting to the biggest cities. To a fair degree, this is a return to the higher rural density that prevailed before World War II, especially in the South and Northeast. It is significant that the greatest growth is in forested or arid areas, rather than in rich farming areas.

My research findings differ from the views of several contributors to this book and Irving Kristol on one point. It is not true that rural settlers, even those around large cities, are necessarily the affluent and sophisticated, who will demand urban-quality services. This phenomenon tends to occur around the very largest metropolises—New York, Chicago, Los Angeles, Philadelphia, and San Francisco. But in many areas, including much of the Northwest, the majority of settlers are below average in income, though rather educated, and have anticity, antitechnology, and rather traditional values; they are people who often do come to fit in with the local economy and culture. Lower housing costs and less stringent regulation have been found to be important incentives for exurban settlement. In my studies of migration to small communities in Washington, Oregon, Montana, and Idaho, I was amazed about the speed with which people adopted local values, including identification with the interests of local employers, who in these communities were mainly in the wood products sector. Americans are not homogeneously urban and acquisitive in culture, despite TV. Rather there is a strong drive for a simpler life, closer to nature—a desire that will continue to fuel much rural settlement.

It is true that urban and recreational demands can adversely affect forestry and this impact will perhaps even increase, if the population continues to decentralize. Do these impacts justify social intervention in the land market? Courts, reflecting the Constitution and our cultural traditions, recognize that individual use of property may inflict external damages on the property of others, and on the well-being of society. Thus societies legitimately regulate individual use of land, as by zoning, so as to minimize such adverse externalities, and—it used to be hoped—conflict about them.

The tough question, at every place and time, is the degree and kind of intervention that is both needed and legitimate. Urban expansion, as it occurred historically, obviously affected preexisting uses, but until recently these effects were deemed benign, since they increased the value and utility of the land. The contemporary justification for containment of urbanization, via farm or forest zoning or purchase of development rights, must rest on the idea that the easily measured increase in the value of land, productivity, and tax yields from urbanization is more than balanced by less easily measured, long-term costs—such as the provision of public services, longer distance commuting, added costs for farming and forestry, and reduced productivity due to lack of open space. Unfortunately there has been no valid, comprehensive analysis—perhaps because studies have been done by those who believe in development or those who oppose it—that really measures the net costs and benefits of greater or lesser degrees of land use control. Dispersed settlement, which is pejoratively called 'scatteration' or 'sprawl,' does not, by its nature, inflict anywhere near the social or environmental costs that would justify market intervention to prevent it.

Thus strategies to prevent rural settlement and to contain cities by timber and agricultural zoning, or by imposing an urban growth perimeter, are probably not wise—that is, these simple and most geographic of planning tools are probably the worst ones to use, because they indiscriminately attack a settlement form, rather than individuals or groups that may create social costs. Such measures tend to restrict the land supply, raise costs, and lower real income; that is, taking pressure and demand off the few reduces the productivity of the many. But planning in the form of influencing the location and pattern of rural fringe settlement and recreational use may well be effective. It is far wiser to accept the demand for rural settlement and to encourage or channel it into rural sectors of more fragmented ownership and away from the most productive farm or forest land, as well as to encourage less disruptive settlement forms, via planned unit developments and performance zoning. As Healy observes, it may also be possible to encourage consolidation of particularly productive holdings through exchange of parcels, as I believe is being considered by the Department of Natural Resources in Washington State. There is probably no escape, then, from somewhat lessened forest productivity as a result of fragmentation, taxes, and managerial restrictions in the vicinity of cities because of the powerful demand for rural settlement. This is the price—really rather small—our society pays for a reasonably open land market and freedom of mobility. The total level of human satisfaction and economic productivity is quite evidently much greater than if we had imposed simply ordered land use patterns and prevented change.

4

TRADITION VERSUS CHANGE IN FOREST LAND USE

Thomas R. Waggener

Robert Healy has presented a concise overview of where the United States is in the early 1980s with respect to urbanization and forest land use. He has pointed out several characteristics of urbanization and land use that will be helpful in understanding the land management problems which urbanizing forest areas will face during the balance of this decade. As resource managers, we must develop an agenda capable of recognizing these characteristics in land management. In quick summary, important points presented include the following five.

1. Urbanization is both geographical and cultural. Foresters have tended to emphasize the geographic dimension of land use and have concentrated on acres—acres presumably devoted to commercial timber production. Urbanization is seen in this context as a nibbling away of acres at the margin as urban uses invade the geographic turf of forestry. A battle line is drawn which becomes the interface—a boundary presumably not to be crossed without strong objection. The lines on the map are set, and they define the margin between land uses. Change is viewed with alarm.

The cultural dimension of urbanization is just gaining recognition within the forestry fraternity, although the implications are hardly perceived. Culture here represents a way of thinking about resources. Forestry has historically represented commodity production. Urbanization has brought about a change. While forest products are obviously important to our modern society, so are alternative goods and services derived from the forest. One does not need to be in the urban fringe to see the impact of this cultural influence. If the objective is indeed to provide a desired mix of outputs, the cultural influence on the redefinition of the market basket of forest outputs must reflect new demands. This leads us to Healy's second important point.

2. Urbanization has led to new and different demands for resource outputs—some of which, we might add, fall outside the traditional agenda of forest land managers. Forestry has traditionally stressed the production technology for wood products. As long as this was the principal desired use of the forest, such an approach was adequate. But competency in wood production does not ensure equal competency in other forms of land management and production. The professional credibility of forestry ultimately hinges on the response to a changing mix of demands and a demonstrated competency (and willingness) to maximize the aggregate output value, dollar and otherwise, of forest use rather than vigorously defend forestry as meaning maximum timber production.

3. 'Replacement demands' at the margin between land uses have created a much larger gross shift in land use than is revealed by traditional 'net' land use change

statistics. Healy has correctly pointed out that net land adjustment statistics overlook the larger total adjustments. An acre of forest land may be shifted to highway use in Washington while another acre finds its way into forest use from old agricultural land in the South. From a national perspective, the net effect is 'no change.' Such regional changes can alter the pattern of forestry without appearing significant in total net change. It is also important to note that replacement demand or, to coin another term, 'land use bumping' depends not so much on the change in forest values as on the values for other uses that may be two or more steps removed from direct competition for forest lands. Industrial or business uses in the urban core will 'bump' residential neighborhoods, which in turn 'bump' residential use into the suburbs. This in turn 'bumps' crop agriculture into pasture lands, which then perhaps 'bump' forest use. Agriculture, in this case, is really not the basic cause of forest displacement. We simply need to look more carefully at the dynamics of such changes if we are to gain a proper perspective of the underlying causes rather than simply observe the symptoms.

4. Juxtaposition of nonindustrial ownerships, traditional forest ownerships, and parcellation has created new problems of economic scale and management for conventional forest managers. It can no longer be a case of business (or forest management) as usual. Furthermore, the movement of urbanites into the woods brings the urban culture (and values) into direct conflict with the rural residents' values, which traditionally have accepted (if not appreciated) forestry as a way of life and a means of community income. Many new rural residents bring their incomes with them and are not dependent on the rural resource sectors to provide a day's pay.

This juxtaposition of ownerships and parcellation has also changed the definition of the urban/forest interface. This interface is no longer a semidefined ring around the urban area. We now have a 'diffused' or dispersed interface which more correctly appears as an overlay on the total land use map.

Foresters have long discussed the nature of economies of scale in forest management. Whether the 'minimum' economic unit is 200 acres, 20 acres, or (as proposed in the state of Washington) 6 acres, we must also recognize that the economic scale for nontimber uses may also be critical. Aesthetic and recreational uses may be influenced by the proximity to commercial timber operations. One only needs to examine the Sunday want ads for rural recreational tracts to gain some appreciation of this fact. A recent look at the classified ads in the *Seattle Times*, for example, revealed that fully one-third of the small parcels listed for sale included some direct reference to abutting forest lands. Phrases such as 'adjoins National Forest' were common. One can be quite confident that the seller does not intend that the potential buyer envision bulldozers or a fresh clearcut next door. To a great extent, the aesthetic value generated by the next-door forest is capitalized into these small tracts. It is easy to see why the purchasers of such tracts are turning to the courts to defend 'their rights' when the unsuspecting forest owner attempts to follow through on long planned forestry operations.

5. The increasing importance of the 'land constraint' of land availability will impose incentives for more intensive forest management, and this 'most worrisome impact' will result in sociopolitical interference with forest management and timber harvest.

Forestry has traditionally been placed at the bottom of the land use hierarchy. It is seen as the use of last resort, unable to hold its own against the onslaught of 'higher' land uses commanding higher prices for land. Forestry is thus viewed as being on the defensive—attempting to hold off the encroaching land use margin. If Healy is correct, the 'land constraint' and the resultant incentive for intensified management may lead to increased land values for forestry in the future, at least where the intensified practices are economically feasible. To the extent that this occurs, forestry will flex its muscles and fight to hold on to and reclaim lands previously released to other uses. Politically, we can expect forestry to stay and fight for its existence at the interface rather than quietly concede.

Increasing land values will also create pressure to improve management of all forest land, public and private. While we were land rich, we could 'afford' to manage for one-third or one-half of site productivity. This simply reflected the economics of management where the resource inputs necessary for intensification were more valuable than land.

Environmental interests have already suggested that we could have equal (or more) timber by managing high-site lands more fully while at the same time releasing 'marginal' lands for environmentally desired uses. One must be careful, however, in looking at physical productivity alone. The U.S. Forest Service recently undertook a 'trade-off study' which revealed that the real constraint was not forest productive potential but the legal and social restrictions already placed on timber management.

These five points raised by Healy, together with his excellent capsule summary of recent land use trends, permit us to examine why the land use pressures on forests in urbanizing areas appear to be suddenly escalating and, more important, what policy issues are likely to emerge in the years ahead.

Although the economics of land use and allocation will be addressed later in this book, a few insights suggested by Healy's chapter are offered here. First, the United States was essentially an 'empty' country, and the settlement (hence land use pattern) was a reflection of the economic and political forces motivating our population in seeking the use of land, for well over two hundred years. This process can be described as one largely based on the 'absolute' economic feasibility of land use—the economist's concept of the extensive margin for feasible land use. If land could 'pay its way,' it was taken up under a wide variety of land disposal laws designed to transpose the public domain into private landholdings. Indeed, land within the extensive margin which would have yielded positive returns was frequently 'passed over' in settlement simply because 'better' land was abundantly available with little or no additional effort.

In the case of forestry, this pattern of 'absolute value selection' resulted in prime forest land being claimed in private ownership, with or without the formal requirements for title transfer being satisfied. The resulting pattern of private ownership obviously reflected the prevailing influence of the geographic distribution of population and the natural inherent productivity of the forest (most frequently determined by the standing old-growth inventory).

The introduction of public forest ownership as it evolved just before the turn of the century also followed an 'absolute selection' process, although the lands available for inclusion in the public forests were significantly influenced by the prior selection of lands for transfer to private ownership throughout the earlier century. The national forests as we know them today were essentially a selection from the leftover remaining public domain thought to be the 'best' forest land not yet taken up in private ownership. As such, these land withdrawals were more opportunistic than the result of a rational calculation that those particular acres should forever remain publicly owned forest.

As the availability of unsettled lands was reduced, conflicts of 'relative value' selection became dominant. Land use depended not only on a measure of absolute feasibility but also on a relative priority as the 'highest and best' use. Healy has indicated that forestry has historically played a role as the 'last' residual land use, occupying those sites beyond the extensive margin for other competing demands.

However, the pressures of (1) intensified management, (2) juxtaposition of ownerships, and (3) urbanization into the exurbs have resulted in a clash of values in which forestry is again threatened with displacement. This is partly the result of the increasing value of forests, hence the economic incentive to retain forest use. It also reflects the impact of more intensive management practices, which bring the encroaching public into direct (and often visual) contact with the messy side of forestry.

As long as forestry was practiced beyond the extensive margin of alternative land uses, little conflict existed. The important lesson to be learned here is that forest management, if it is to remain a viable land use, must not only reflect the economics of intensification where it is feasible but also the modifications necessary to accommodate urban neighbors. We can anticipate that forest management textbooks in the future will read quite differently from those in use today. Maximum fiber production (even within the limits of economic constraints) will increasingly be contrasted with reduced emphasis on fiber and increased emphasis on those unmeasurable amenities that are almost impossible for a private forest landowner to capture.

Several outcomes are possible. First, the forest landowner can concede and withdraw to more remote and less visible locations to practice the trade. Second, the owner can dig in and attempt to defend traditional management practices as being within the recognized property rights associated with private landownership. This avenue, it should be noted, runs the risk of legal intervention to redefine property rights. Whatever one's personal views, it is not evident that conventional forestry will prevail.

Another strategy might be to recognize the intrusion of urban values as a new 'cost of doing business.' As objectionable as this may seem to the hard-hat forester, it may be the only feasible strategy for survival at the urban/forest interface. The manager has an incentive to make changes that will minimize the total cost of getting the job done. This may result in modified management rather than conceding forestry use at the interface. The tendency to defend tradition has perhaps short-circuited the search for alternative management techniques.

Finally, it must be realized that commercial forest use is frequently not the only feasible use—or necessarily the highest use—and that land use changes can be socially desirable. This is true even if it means displacing the long-standing commercial forest use of certain lands. Debates about the 'loss of forest land' are not complete without recognition of the trade-offs to be gained from alternative uses. Healy's concept of replacement demand makes a major contribution to our understanding of this dynamic land allocation process.

It is evident that land use allocation is neither fixed nor static. As relative values of resources and resource outputs change, land allocation should also change as owners and managers seek to put lands to the best use. This dynamic process necessarily entails economic friction. Adjustments are never instantaneous or perfect. Delays occur as owners and managers try to assess the signals sent by society. Some of them may make mistakes and misread the signals or react to imperfect information. Finally, the signals themselves may be in error, or at best incomplete. Responses based on such signals may misallocate land resources.

It is easy to react to such institutional friction and declare the dynamics of land allocation undesirable. It would be a simpler world indeed if such changes did not need to occur. Unfortunately, much of our land policy is well intentioned yet designed to avoid the friction of land adjustments rather than to smooth the process. For lands in the private sector, restrictions may be placed on land use changes with the intent of preventing the short-term costs of adjustment. In the public sector, land use may well become legislatively frozen in reaction to economic forces indicating the desirability of change. In either case, the delay or resistance to change will simply widen the differential between values of alternative uses and postpone the necessary adjustments. With such delays, however, the ultimate adjustment may impose substantial costs and controversy. Much is at stake for both existing users and potential users.

Viable land use policy, including forest land use, must remain flexible enough to adapt to changing conditions. Our society wants wood products—but also other goods and services. This is what the debate surrounding the urban/forest interface is really about. How can we best use our valuable land resource to meet the demands of our diverse population. Healy has pointed us in the right direction—toward solutions rather than debate.

PART II

LEGAL, POLITICAL, AND ADMINISTRATIVE DIMENSIONS

5

PUBLIC AND PRIVATE ALLOCATION OF URBAN FOREST RESOURCES

James L. Huffman

Scientists have a knack for describing social issues in the most clinical terms. The urban/forest interface' is one such description. I certainly have no quarrel with the title of this symposium; indeed I can think of no better language to communicate both the subject matter and the purely objective (scientific) approach evidenced by the program. It will be quickly evident that I have adopted the term with unexcelled enthusiasm as a convenient shorthand for the description of a complex problem. However, I must confess at the outset that unlike the foresters, economists, sociologists, and most others in attendance, I am not a scientist. There was a time when American legal scholars, probably in search of academic respectability, were generally in agreement about making a science of legal study. Although the influences of that movement still pervade our educational and intellectual enterprise, I have no hesitancy in describing my field of study as nonscientific. Rather, I would describe it as both technical and philosophical.

We lawyers are engaged in the business of designing, interpreting, and applying tools intended to influence human behavior. We are thus unavoidably caught up in the philosophical problems of defining desirable behavior. Of course, lawyers are heavily dependent on the sciences, and some of us are more than willing to dabble in the realms of science as we pursue our lawyerly tasks. But unlike the scientist, the lawyer does not pursue truth in any fundamental sense. Our pursuit is of ends which are defined by human values. I will offer my assessment of what the law is on subjects relevant to this symposium, occasionally with a high degree of confidence, but I will also suggest what legal mechanisms I believe will result in the realization of particular values. Science is essential to evaluating the likely effectiveness of recommended legal approaches, but I do not believe science can answer the most fundamental question of all, 'What should we do?' Thus I would amend the title of this symposium ever so slightly to read, 'The Urban/Forest Interface: Land Use and Forest Resource *Choices* in a Changing Environment.' No doubt my amendment only underscores what is implicit to all who understand the issues we are addressing, but the basic thrust of my approach makes the emphasis on choice important.

THE SITUATION

In many parts of the United States, the only trees visible to urban dwellers are those they or their predecessors planted. In other regions, particularly the Pacific Northwest, urban areas have grown up in the trees. Cities like Seattle and Portland owe much of their character, aesthetic and cultural as well as economic, to the forests they inhabit. As these and many other cities have grown, the surrounding forests have receded. Of course, urban growth is not the only, or the most significant, cause of the conversion of forest land to nonforest uses. The single most important cause has been the expansion of agricultural lands, a conversion that has been taking place since the earliest European settlement of the continent.

As recently as 1973, Professor Lee James told the annual meeting of the Society of American Foresters that 'the rapid advance of urbanization is not ... a direct threat to forestry' (James 1974). A 1980 publication of the Environmental Law Institute on the regulation of lands with 'special natural resource values' treats the conversion of forestry areas as an exclusively rural problem (Kusler 1980). Indeed, an exhaustive search of the legal literature reveals a single article on the subject of forest land preservation, in which the author notes that the 'problem of shrinking forestlands ... has thus far received little attention.' Presumably this conference is evidence that forest land conversion to urban uses is beginning to attract some attention.

The issue of forest land conversion is a logical outgrowth of the long-standing concern over the conversion of agricultural lands to urban uses. However, land use conversion is not the only, or the most important, issue raised by the urban/forest interface. The methods of forest management, particularly those associated with timber harvesting, have significant impacts on nearby urban dwellers. Air, water, and noise pollution, as well as the alteration of the urban dweller's vista of nature, are concerns for those who live in the city. Forest managers face reciprocal costs as a consequence of their proximity to large populations. Municipalities extend their boundaries into the country to include new urban development, and counties find themselves governing urban as well as rural areas. For both types of local government, this mixture of urban and forest lands creates new issues of land use regulation, as well as public service provision and property taxation.

Other participants in this symposium will examine in depth the economic, political, and social reasons for and implications of the various aspects of the urban/forest interface. My assignment is to discuss the legal and constitutional aspects. The following sections will examine the constitutional bases for governmental action, the constitutional limits on governmental action, other legal restrictions on governmental action, and existing governmental authority at the local, state, and federal levels. In describing existing state and local authority, I will focus on Washington, Oregon, and California. Having thus performed the technical portion of the lawyer's role, I will address two broad policy issues that may take me beyond my assigned responsibilities but will permit me to perform the philosophical function I claim as part of the business of lawyering. The policy issues I will discuss are whether or not, or in what circumstances, governmental action is appropriate to cope with the problems that may exist at the urban/forest interface, and to the extent that governmental action is justified, the level of government and the methods most appropriate to the task.

THE CONSTITUTIONAL BASES OF GOVERNMENTAL ACTION

Federal, state, and local governments have expansive authority under the federal and state constitutions to regulate and otherwise influence the use of both urban and forest lands. The commerce clause (Art. I, § 8, cl. 3), the property clauses (Art. I, § 8, cl. 17, and Art. IV, § 3, cl. 2), the general welfare clause and the taxing and spending powers (Art. I, § 8, cl. 1) combine to give the federal government almost unlimited regulatory powers. The police powers of the states, although circumscribed by both the enumeration and exercise of federal powers, are broad, particularly in the area of land use regulation. (Land use regulation has been an important part of the police power of state and local government since the Supreme Court decision in Euclid v. Ambler Realty, 272 U.S. 365 [1926].) Local governments, which derive their constitutional authority from the states, have generally had similarly few constraints on their authority to regulate land uses. Taken together, federal, state, and local governments are in no way seriously limited in their ability to regulate the problems associated with the urban/forest interface by a lack of affirmative constitutional authority. Of course, the existence of affirmative constitutional authority does not usually require governments to exercise that authority, as the discussion of existing regulations will demonstrate.

American governments are not therefore free to pursue every possible regulatory end, nor are they free to employ every conceivable means of regulation. The Fifth Amendment to the United States Constitution prohibits deprivations of property without due process of law, and requires just compensation where private property is taken for public purposes. The Supreme Court's interpretation of the latter restriction has evolved from a distinction between the police and eminent domain powers of government to ad hoc assessment of the impact of a particular regulation upon the value of affected property rights. (See Pennsylvania Coal Co. v. Mahon, 260 U.S. 393 [1922] and Penn Central Transportation Co. v. City of New York, 98 Sup. 2646 [1978].) At no time in the last half century has the Court's interpretation of the takings clause imposed serious limitations on governmental regulation of private property. The due process requirement in no way limits the objectives of regulation; rather it requires that governments comply with constitutional, legislative, and administrative rules of procedure.

The substantive purposes of regulatory action are limited by the First Amendment guarantees of freedom of speech and religion, and by the Fourteenth Amendment's equal protection clause. The Fourteenth Amendment, through both the due process and equal protection clauses, places procedural limitations on the states and assures that the substantive limits of the First Amendment will apply to the states as well as the federal government. (For a discussion of the equal protection and First Amendment issues see Merriam 1981; for a discussion of the due process issues see Smith 1981.) Although it would be a serious mistake to ignore the important ways in which the federal and state constitutions limit governmental power, within the realm of policy options seriously considered by anyone involved in land use regulation and forest management, constitutional law presents few obstacles even to the most ambitious advocates of public action. Of course, advocates of all policy positions might well propose constitutional amendments or seek constitutional reinterpretations as means of pursuing their policy goals.

The constitutional issues are not limited to the definition of the power of governments in relation to private groups and individuals. The power of governments in relation to one another is centrally a constitutional issue, and is of particular importance today in light of the debates stimulated by the Reagan administration's proposals for a 'new federalism.' (See *Symposium on Federalism and the Environment*, 12 ENV. LAW #4 [1982].) The states have played a dominant role in land use regulation since the Supreme Court first approved zoning as a legitimate exercise of the police power (Euclid v. Ambler Realty). Whether or not the federal government might constitutionally preempt some or most of the land use regulation authority is an issue often debated, but the courts have not had the opportunity to consider it because of the failure of Congress to enact national land use legislation. This does not mean that the federal government has little influence on land use, particularly on forest lands, a significant proportion of which are owned by the federal government and managed pursuant to the property clauses of the Constitution. Pursuant to the commerce and general welfare clauses and the taxing and spending powers, the federal government also regulates many land uses indirectly.

The constitutionally prescribed line between the powers of the federal and state governments is not clearly defined, although it is clear from the last four decades of Supreme Court decisions that the powers of the federal government are both expansive and expandable. To the extent that the Constitution is interpreted to limit the exercise of certain powers to a single level of government or that practicality necessitates a single authority, the expansion of federal power inevitably means the contraction of state power. (The distribution of powers between the federal and state governments is based on the notion that a few powers are exclusive to the federal government, many powers are concurrent to federal and state governments, and those powers not delegated to the federal government are reserved to the states. With respect to concurrent powers, action by the federal government may preclude further state action in that area under the doctrine of preemption, which is based on the supremacy clause of Article VI.) For example, if the Congress were to adopt a national land use regulation law, the potential scope of state and local land use regulatory power would be diminished in proportion to the power asserted by the federal government. Opponents of federal land use regulation could certainly challenge a federal law as unconstitutional, but without a dramatic shift in Supreme Court interpretations of federal power, there is little reason to think a federal land use law would be found unconstitutional. Although the experience of the post-New Deal decades would suggest an inevitable expansion of federal power at the expense of state power, the anti-big government thrust of the Reagan administration should make it clear that the federalism issue is far more political than it is constitutional. Because so many governmental powers are shared by the federal and state governments, in the absence of federal preemption the extent of potential state authority is to a large degree a function of political decisions at the federal level on what federal constitutional authority will in fact be exercised.

One area of conflict over the relative powers of the federal and state governments is particularly important to the urban/forest interface. Some states and local governments have claimed the authority to impose land use regulations on federal lands that fall within their jurisdiction. Federal authorities have responded by claiming that under the property clause of Article IV, they have exclusive management authority over federally owned lands. Furthermore, federal officials have claimed limited authority over private lands adjacent to federal lands to the extent that the use of those private lands affects the federal ability to fulfill its management responsibilities. Thus, the states are seeking to extend their traditional land regulation authority to federally owned lands while the federal government seeks to extend its traditional land management authority to private lands.

There has never been any question that the states have jurisdiction over certain activities on federal lands through the application of state civil and criminal laws. States have traditionally regulated wildlife whether on federal, state, or private lands; and, despite some federal wildlife regulation, the states continue to exercise that authority. But the ability of states to extend traditional land use regulations to federal lands is far less certain. The Supreme Court has stated that '[a]bsent consent or cession a State undoubtedly retains jurisdiction over federal lands within its territory, but Congress equally surely retains the power to enact legislation respecting those lands pursuant to the Property Clause' (Kleppe v. New Mexico, 426 U.S. 529, 543 [1976]). When Congress does act pursuant to the property clause, the supremacy clause requires that all inconsistent state regulation give way to the federal exercise of authority. Because most land use regulation deals with the very issues that are central to any federal land management program, it would appear that state land regulation authority on federal lands will be preempted in significant part. At the same time, state authority over nonfederal lands will be restricted to whatever extent the federal government finds it necessary to regulate those nonfederal lands in order to fulfill its management responsibilities on federal lands.

Within very broad constitutional parameters, both federal and state governments can pursue many different approaches to the allocation and management of lands at the urban/forest interface. All governments are constrained to a limited degree by constitutional protections of private rights, and the state and local governments are constitutionally limited by political decisions at the federal level that define the extent of federal involvement. It may be argued that the Constitution should be amended or interpreted to play a bigger role in defining the extent and relative distribution of governmental authority over land use decisions. Some have argued for a private constitutional right to be free from governmental interference in the use of private lands, while others have argued for a private constitutional right to demand such intervention. Similarly, disparate arguments have been made for a larger constitutional role in the allocation of authority among governments in the federal system. However, existing understandings of the relevant constitutional provisions leave government with wide-ranging options. The following sections examine how the federal and state governments have pursued those options.

FEDERAL AUTHORITY AT THE URBAN/FOREST INTERFACE

To the extent that the federal government owns forest lands that have an impact on or are affected by urban development, the federal government has a large role to play. The Forest Service and the Bureau of Land Management are the two principal federal agencies with management authority over any forest lands that might be of concern. Both agencies function pursuant to a long series of statutes that define their public policy objectives and the procedures they are to employ in pursuit of those objectives. The Forest Service, which functioned for decades under the vague guidelines of the Organic Act of 1897 (16 U.S.C. §§ 473-81), is now regulated by the Multiple-Use, Sustained-Yield Act of 1960 (16 U.S.C. §§ 528-31), and the Forest and Rangeland Renewable Resources Planning Act of 1974, which was amended by the National Forest Management Act of 1976 (16 U.S.C. §§ 1601 *et seq.*). Taken together, these statutes define the purposes (not including urban development) for which the national forests are to be managed and the process to be used in making the difficult choices among the permissible alternatives. The Bureau of Land Management, which was formed in 1946 by a merger of the Department of Interior Grazing Service and the General Land Office, is governed by the Taylor Grazing Act (43 U.S.C. §§ 315 *et seq.*), and, more important, by the Federal Land Policy and Management Act of 1976 (43 U.S.C. §§ 1701-82). The latter statute defines a multiple use policy not unlike the one that guides the Forest Service; however, the BLM lands and their historical uses dictate a very different use allocation from that which occurs on national forest lands. Small amounts of federal forest lands in urban areas are administered by the National Park Service and the Fish and Wildlife Service.

The most important aspect of federal authority over these urban forest lands, aside from the fact of federal ownership, is that the governing statutes define a scope of permissible uses which excludes most traditional urban land uses. With the exceptions of recreation and the relatively recent demand for open space in urban areas, federal lands cannot be managed for what would be considered urban purposes. In some cases, this has resulted in an urban/forest interface which is a visible wall of trees on the one side and urban development on the other. Of course, the relationships between urban areas and forest lands are more complex than the boundary dividing the trees from the buildings and asphalt, but this limited use aspect of federal land management is an important constraint on the policy alternatives in areas of impact between federal forest lands and urban development. Two qualifications should be added to underscore that the constraints of federal ownership are not central to policy decisions in the context of the urban/forest interface. First, the federal government does have authority under several statutes to exchange federal for private lands (16 U.S.C. § 516, § 485 and 43 U.S.C. § 1716), and, as Secretary of

Interior Watt reminds us, the federal government also has authority to sell federal lands. (Article IV, § 3, cl. 2 of the Constitution grants Congress the 'Power to dispose of and make all needful Rules and Regulations respecting the Territory or other Property belonging to the United States.') Second, and probably most important, federal ownership of forest lands in or near urban areas is relatively insignificant. The vast majority of these lands are in private ownership.

The fact that we are primarily concerned with private lands, and some lands owned by state and local governments, does not mean that we can ignore the federal government except where it manages its own lands. The federal government plays a major role at the urban/forest interface through various regulations of the external impacts of both urban and forest land uses. Although there is no federal legislation that specifically regulates forest land uses, several federal statutes may be applicable in particular situations. The National Environmental Policy Act requires an environmental impact statement for 'federal actions significantly affecting the quality of the human environment' (42 U.S.C. §§ 4321 *et seq.*), a requirement that has been found to apply to private and state developments dependent on federal financing or approval. The federal Clean Air Act (42 U.S.C. §§ 7401 *et seq.*) and the Water Pollution Control Act (33 U.S.C. §§ 1251 *et seq.*) have broad application to polluting activities, including those resulting from forest management and harvesting and from the manufacture of timber products. Not only do these two acts regulate polluting activities where they happen to occur, but they 'can be expected to have a major impact on land development patterns through the influence they exert on the location of [air and] water pollution sources' (Mandelker 1976). In Oregon, Washington, and California, the National Coastal Zone Management Act is another source of significant federal authority affecting the use of forest lands in urban areas (16 U.S.C. §§ 1451 *et seq.*). Other federal legislation provides further federal regulatory authority, but suffice it to say that the national government has substantial means for exercising direct and indirect controls over the use of forest and urban lands. Any comprehensive land management scheme for these lands must be designed to recognize this broad federal authority.

STATE AND LOCAL AUTHORITY AT THE URBAN/FOREST INTERFACE

State and local governments have primary authority for land use regulation. In addition, they have other regulatory powers that can affect forest and urban land uses. Finally, they can have important impacts through their taxing and spending powers. What follows is a brief review of the relevant laws of California, Oregon, and Washington.

California

Land use regulation in California is largely the responsibility of local government, although the Governor's Office of Planning and Research is involved in the formulation, evaluation, and updating of long-range land use goals and policies (Gov't. Code § 65035). Local governments are required to develop a general plan for the land development in their jurisdictions. The plan must include elements that address the following topics: transportation circulation, resource conservation, open space, seismic safety, noise, scenic highways, safety, and housing (Gov't. Code § 65302). Cities and counties are required to adopt zoning regulations consistent with the local plan, although the vague definition of the state-prescribed planning 'elements' makes compliance difficult to assess. Of particular significance to the development of lands on the urban fringe is the Subdivision Map Act, which controls developments containing five or more parcels (Gov't. Code §§ 66410-499.58). Local governments are not required, however, to consider extrajurisdictional impacts under most of the land use elements of the general plan. The open space element refers specifically to preservation within political jurisdictions (Gov't. Code § 65302), and the conservation element does not require extraterritorial considerations except on water programs (Gov't. Code § 65302d). No area-wide coordination is required among local governments (Gov't. Code § 65302a), and although California has twenty-five Councils of Governments (COGs), they are poorly funded and lack the expertise and authority to make effective regional decisions.

California's Forest Practice Act (Pub. Res. Code §§ 4511-4628) is considered by many to be the strictest and most far-reaching forest practice law adopted by any state. The law is intended to protect all forest resources, including timber, recreation, wildlife and fisheries, and watershed, although existing regulations focus almost exclusively on the regulation of timber harvesting activities. Prior to harvest, an operator must submit a timber harvesting plan, including a description of the land, methods to be used, and other harvesting data. Harvesting may not proceed until the plan has been approved, and on-site inspection may occur at any time. The law applies to all nonfederal public and private lands except for the harvest of Christmas trees and fuelwood and the removal of dead or diseased trees.

Under the California Forest Practice Act, private timber owners must apply for a permit to convert timberland to nontimber uses when such land lies within a timber production zone (TPZ). Permits may be granted by the director of the Department of Forestry only if (1) the conversion would be in the public interest, (2) the conversion would not adversely affect timber-growing or open space land, and (3) the soils, slopes, and watershed conditions are suitable for the proposed new use (Klein 1980). Conversions of less than three acres are excepted from the act's controls. Prior to the adoption of the Forest Practice Act in 1973, forest land conversion was a common means of avoiding reforestation requirements. Thus a central purpose of the new permit system is to assure that conversion actually occurs where there is no reforestation.

In 1976 the California legislature adopted the Forest Taxation Reform Act, which was intended to encourage the retention of timberlands by dealing with those aspects of traditional property taxation that were believed to encourage timberland conversion (Gov't. Code §§ 51100-51150). The perceived tendency of encroaching urban development to push up tax assessments based on a 'highest and best use' measure of value has led to the adoption of a stumpage tax levied at the time of harvest. A central feature of the Forest Taxation Reform Act is the requirement that cities and counties zone all timberlands within their jurisdiction. Private landholders may petition to have their land excluded from a TPZ, and must demonstrate that inclusion of their land within the zone is not in the public interest (Gov't. Code § 51129a). Land immediately removed from a TPZ is subject to the assessment of a recoupment fee designed to repay the state for taxation deferred during the parcel's inclusion in the TPZ (Gov't. Code § 51142a). Land included in a TPZ can only be rezoned at the end of a ten-year period, except when rezoning is at the initiative of the local government.

In 1970, California became the first state to adopt legislation modeled on the National Environmental Policy Act. The California Environmental Quality Act requires an environmental impact report by local governments for 'any project they intend to carry out which may have a significant effect on the environment' (Pub. Res. Code §§ 21000-21176). After a California Supreme Court decision applying the law to private activities regulated by a state agency (Friends of Mammoth v. Board of Supervisors, 500 P.2d 1360; modified 502 P.2d 1049 [1972]), a 1972 amendment extended the environmental impact report requirement to all discretionary projects proposed to be carried out or approved by public agencies. The Environmental Quality Act specifically excludes timber harvesting activities from its coverage on the assumption that the timber harvesting plan of the Forest Practice Act serves the same purpose. However, it appears that environmental impact reports may be required prior to the issuance of a permit to convert private timberlands to some other use.

Oregon

The much praised and regularly challenged Land Conservation Development Act is the basis for a system of statewide planning in Oregon (§§ 197.005-.430). The act created the Land Conservation and Development Commission, which is responsible for the promulgation of land use goals and guidelines and has the power to require local governments to develop comprehensive plans that comply with the promulgated goals and land use regulation procedures. The statewide goals have the purpose of defining the subject areas that local plans must address and providing minimum standards that local zoning regulations and comprehensive plan changes must satisfy. The system is designed so that the Land Conservation and Development Commission makes broad policy decisions while all site-specific decisions are made locally. Site-specific decisions are subject to review of the Land Use Board of Appeals and the state court of appeals.

Goal 4 requires the conservation of forest lands for forest uses. Goal 5 provides for the conservation of open space and the protection of natural and scenic resources. These goals require that a local comprehensive plan provide for the conservation of forest lands, both for timber and nontimber purposes, and thus limit the private conversion of forest lands to urban uses. Goal 14 requires that the local comprehensive plans are to 'provide for an orderly and efficient transition from rural to urban land use' (LCDC 1978), making it clear that Goals 4 and 5 are not intended to be absolute prohibitions on forest land conversion. To the extent that the use of lands adjacent to forest lands affects forest uses, the adjacent lands must be zoned to minimize the detrimental impacts. In practice the state Department of Forestry conducts studies for county and local planning agencies to help determine whether or not conversion should be allowed. The department ranks forest land on a productivity scale from 1 to 5, and usually recommends that any average (3) or better (1 or 2) forest land in western Oregon be continued in that use. The final decision is up to the local planning agency as long as its decisions conform to the LCDC goals and guidelines.

In 1971, the Oregon legislature revised its forest practices law to provide for a comprehensive approach that considers all forest uses (§§ 527.610-.990). The Forest Practices Act authorizes the state Board of Forestry to prescribe rules and regulations concerning reforestation, road construction, harvesting, chemical applications, slash disposal, and water quality and watershed preservation. It applies to all nonfederal public and private forest lands. Timber operators are required to notify the state forester before commencing operations, and the Department of Forestry in turn sends a copy of rules with which the operator is expected to comply. However, the act expressly disclaims any grant of authority over the conversion of forest lands, thus leaving that issue to be dealt with solely in the context of the state land use planning system. The only role for the Department of Forestry with respect to the conversion of forest lands is the making of a determination of actual conversion in the case of those lands where exemption from reforestation requirements is sought (Admin. Reg. § 627-24-103).

Oregon no longer employs an ad valorem property tax on standing timber. In western Oregon, a severance or yield tax of 6.5 percent is levied at the time trees are harvested (§ 321.247[9]). In eastern Oregon, the rate is 5 percent on the 'immediate harvest value' of all timber, whether or not actually cut (§§ 321.405-.520). The severance tax approach was adopted in an effort to remove disincentives to timber production believed to result from traditional property taxation methods. Forest lands may be designated as such for property taxation purposes, thus avoiding highest and best use assessment which would push tax bills up in those areas on the urban fringe. Property classified as forest land is liable for the payment of some additional taxes if and when the forest land classification is removed.

In 1979, the Oregon legislature passed the Woodland Management Act to encourage investment in forest land maintenance and to reduce disincentives resulting from ad valorem property taxation (§§ 526.450-.457). The act is intended to provide tax relief during timber growth periods, promote programs to provide credit on young stands and encourage harvesting of mature stands, promote the planting of new forests on cutover or unproductive private lands, and protect the public interest by assuring a continuous production of forest products. Aside from making provision for a forest tree seed bank, the act is exclusively a tax measure designed to provide economic incentives for forest land management by small private landowners. A 10 percent state income tax credit is given owners of between 10 and 500 acres who convert cutover or unproductive land to the production of commercial timber stands (§§ 316.094, 317.102, 318.030).

Washington

Under Washington law, cities and counties are empowered to develop comprehensive plans, but they are not required to do so. Despite numerous efforts in the legislature, a statewide comprehensive planning system does not exist in Washington. A 1977 bill would have created a Land Conservation and Development Commission similar to Oregon's (S. Bill 65), but the legislation did not get the necessary votes. Thus comprehensive planning in Washington conforms to the provisions of the Standard Zoning Enabling Act, which equates a comprehensive plan with the policies of an existing zoning ordinance (Kozesnik v. Montgomery Township, 131 A.2d 1 [N.J. 1957]).

The Washington Forest Practices Act of 1974 represents a compromise among political interests (§§ 76.09.010-.935). It is administered by the Departments of Ecology and Natural Resources with input from the Departments of Fisheries and Game. Rules and regulations are promulgated by the Forest Practices Board with the advice of a Forest Practices Advisory Committee. The regulatory structure is based on a system of classification that prescribes different governmental responses to different types of actions. All classes of activities require compliance with regulations, but in some cases operators may proceed at their discretion while in others a permit or even an impact statement may be required. The law applies to all private forest lands, and on-site inspections by the Department of Natural Resources are authorized.

A primary objective of the Forest Practices Act is the promotion of reforestation. Thus, operators who must apply for permits are required to declare whether or not they intend to convert the land to a noncommercial timber use after harvest. After harvest, the landowner must demonstrate actual conversion within three years or reforestation will be required. However, reforestation may not be required where the land is not capable of supporting a merchantable stand of timber or in areas identified by the regulations as being likely to convert to urban uses within ten years. A state determination that an area is likely to be converted from timber use must conform to local comprehensive plans for the area (§ 76.09.070). In all cases where an applicant indicates that conversion will occur, local zoning authorities have the power to object and a permit may not be issued by the state. The act provides for appeal by any interested party, including the Department of Natural Resources, to the Forest Practices Appeal Board (§ 76.09.220[8]a).

Washington's Forest Taxation Act of 1971 provides for assessment of forest lands in their current use rather than at market value if those lands are 'primarily devoted to and used for growing and harvesting timber' on tracts of 20 acres or larger (§ 84.33.100[1]). Lands are graded on the basis of quality, accessibility, and topography, with assessments varying accordingly. Landowners may petition to have their lands classified as forest lands, and the assessor may petition for removal of lands from classification. Reclassified lands are subject to deferred taxes on the difference between previously assessed and current market values multiplied by the number of years the land benefited from differential assessment up to a maximum of ten years. Under the Open Space Taxation Act, 5- to 20-acre farm woodlots may be classified as 'farm and agricultural land' and larger parcels may be classified as 'timberland' (§§ 84.34.010-.160). When so classified, lands benefit from use-value assessment as agricultural lands.

The Washington legislature has also authorized the purchase of development rights to timberland by cities, counties, or nonprofit nature conservancies (§§ 84.34.210-.250). Counties have the authority to establish a conservation futures fund based on a county-wide property tax levy. Implementation of this program is discretionary with local governments.

The Washington Environmental Policy Act requires state agencies and local governments to consider the extent to which any 'major action' will cause adverse environmental effects in excess of those under existing uses (§ 43.21c *et seq.*). The Washington State Supreme Court has held that the issuance by governmental agencies of permits and other entitlements for private actions requires review under the act (Eastlake Comm. Council v. Roanoke Assn. Inc., 513 P.2d 36 [1973]). Environmental impact statements are required to address, in addition to environmental impacts and alternative courses of action, any irreversible commitments of resources.

WHAT SHOULD WE DO AT THE URBAN/FOREST INTERFACE?

The preceding sections survey much of the existing law at the federal level and in three western states as it relates to the allocation of forest resources on the urban fringe. More could be said about the laws of the particular jurisdictions. For example, the land use authority of local government may be limited by antitrust legislation, and the amenability of local government to antitrust challenges may, in turn, be limited by doctrines of governmental immunity (Netter 1981, Bosselman 1981). The complexity of these potential legal issues is limited only by the creativity of the lawyers who represent the respective interests. Local governments have important influences over land developments through their decisions with respect to the investment of capital resources in public services like roads and water and sewer mains, although their ability to limit growth by these means may be subject to some constitutional limits. Indeed, every aspect of government spending, taxation, and regulation may affect in some way both public and private land use decisions. A treatise on the subject would require an investigation of all these issues, and much greater depth on the topics that have been discussed above. For the purposes of this chapter, however, the foregoing survey will suffice to illustrate the complexity of the legal and constitutional issues and to provide a context for the following discussion of policy.

We cannot escape the task of allocating and distributing scarce resources in a finite world. Societies have adopted an assortment of mechanisms, but the results are never satisfactory to all members of society. For those who are satisfied, a problem disappears; for the others it remains. But for those who remain dissatisfied, there has been a subtle but important shift in the nature of the problem. No longer is the problem to find a social mechanism for resource allocation and distribution; the problem is that the chosen mechanism does not lead to resource allocations and distributions that satisfy the values of all individuals. Those who are dissatisfied will seek to alter or replace the existing social mechanisms so that they deliver more personally satisfactory results. Except in some mythical state of nature, we have never really had the opportunity to address the problems of resource allocation and distribution in the abstract. We have always had to address the issues in the context of existing social mechanisms and existing resource allocations and distributions.

In a society that holds democracy as an ideal form of governmental organization and increasingly looks to government to satisfy the needs and wants of individuals, it is not surprising that situations and circumstances are described as problems, not for the individual but for the society. The question is always what are *we* going to do about *our* problem, not what is the government going to do about the individual's problem or what is the individual going to do about his or her problem. Democratic government exists to serve the public interest, not the interests of individuals, so in seeking action from a democratic government, we must appeal to collective, not individual, problems. As politicians or individuals concerned about how government treats us, we are well advised to deal in problems if action is what we want. But as academics, our first task is to understand physical and social realities. Once we have done that, we will be in a position to advise people on the solution of problems. We will not be led, nor lead others, astray if we remember that a problem exists and is solved only in relation to the standards or values against which an outcome is evaluated.

The situation at the urban/forest interface is well described by Robert Healy, with important contributions from each of the other discussants. I have summarized the legal dimensions of that situation which I believe are in significant part a consequence of viewing as social problems a lot of things that were, in fact, private problems. Although legislation to deal with the 'problem' of forest land conversion and the 'problem' of the external costs of timber harvesting is of recent origin, most of the relevant laws are the product of a multitude of government responses to political pressures. We face a complexity of laws which I have discussed under the subject heads of land use, forest practice, environmental, and tax laws. These laws are administered by an assortment of agencies at several levels of government. Even if we find the articulated purpose of each law to be meritorious, their separate and combined impacts on whatever objectives we may have for our urban forest lands are a matter of considerable uncertainty.

How will a particular landowner's decision about the use of his land be affected by the following combination of laws? His county's comprehensive plan includes his parcel in an area to be zoned for commercial forests. The county has not zoned his land exclusively for timber production, but the state's land use legislation requires county zoning to conform to the comprehensive plan unless the public interest justifies a variance. The nearby city has recently extended its water and sewer systems to the adjacent parcel of land that is in the same comprehensive plan designation, but for which development approval was granted prior to the adoption of the comprehensive plan. The state forest practices act imposes serious restrictions on timber harvesting done in urban areas and requires reforestation within two years of harvest. Christmas tree growers are exempt from the provisions of the forest practices act. The forest practices act also prohibits conversion of forest lands to other uses unless such conversion conforms to the local comprehensive plan or if the department of forestry determines that the lands are inappropriately designated for commercial timber production. The state environmental policy act requires state agencies that approve significant development projects to file an environmental impact statement, but timber harvesting is exempt because of the similar requirements of the forest practices act. The land is taxable on the basis of its value as commercial timberland, although deferred tax payments are required if the land is converted to nonforest uses. The state assesses a yield tax on timber harvested based on the value of the timber. The land is adjacent to a national wildlife refuge and is located upstream on a river from which the city diverts a portion of its water supply. The landowner's lawyer advises him that the state and federal constitutions prohibit taking of private property without just compensation, but acknowledges that the courts have been very restrictive in defining what constitutes a taking.

Complex as the foregoing hypothetical case would be, it is only a sampling of the laws and legal institutions that might be relevant to a single decision about the allocation of private forest land on the urban fringe. The private owner will act on his best judgment of how his action, given all the relevant laws, will serve his interests. It will not be an easy decision, nor will the outcome be predictable with a high degree of certainty. Of course, since our laws are motivated by a desire to serve the public interest, we are less concerned about the impacts on the individual landowner than we are about those on the public interest. But our public actions are constrained by the same uncertainties, with the added difficulty of having to determine the public interest. The private decision maker is at least certain about what he or she would like to accomplish.

As we multiply our efforts to alter the allocational impacts of private decisions, whether through public ownership or regulation of private ownership, we multiply the opportunities for private interests to employ the power of the state to achieve favorable redistributions of wealth. Indeed, this prospect is a major cause of expanding public intervention. Generally that intervention is done in the name of better resource allocation, but the distributional consequences of particular governmental actions are far clearer than are the allocational consequences. It is a difficult empirical and theoretical problem to determine whether society is better off as a result of prohibiting development of lands zoned for commercial timber production. It is a far simpler matter to determine that those who would prefer to see trees in the distance, rather than houses, are better off as a result of the regulation.

I do not intend to suggest that we should kick the government out of the decision making on the uses of urban forest lands. I have no doubt that certain private actions will result in a net loss to society as a whole, but I believe it is fundamentally important that we have a clear understanding of why those losses occur and what we can reasonably expect to do about it. A few years ago Reuben Plantico and I ventured far into the realm of economics to construct a theory for land use regulation in the context of Oregon's comprehensive statewide system (Huffman and Plantico 1979). Our argument, borrowed entirely from the economists who have articulated it far better than my knowledge permits, was that land use regulation should be designed to remedy the failings of the marketplace as an efficient allocator of resources. This argument took us through the familiar territory of external costs, public goods, and free riders to the conclusion that Oregon's land regulation system could not be justified in the name of economic efficiency. That Oregon persists, despite our criticism, does not surprise me, since I believe the real motivations for land use regulation have little to do with efficient resource allocation and a lot to do with wealth redistribution. A case for land use regulation to redistribute wealth might be articulated, but it will be difficult to persuade those who pay the costs that it is in the public interest to transfer their wealth to others.

I will not attempt to reconstruct the case for public intervention to achieve efficient allocation of the resources at the urban/forest interface. Suffice it to say that certain land uses, perhaps recreation or wildlife habitat, may be public goods (external benefits) which the market will not provide. Other land uses may lead to external costs in the form of air or water pollution which if internalized would result in different resource allocations. But if these inefficiencies exist, a fact that we may be prevented from determining given the impacts of existing regulation, it is unlikely that efficient allocation will result from a system of overlapping and intersecting regulatory initiatives. What will happen is that private interests will be forced to pursue those interests in the political arena, where, unlike the market, you can sometimes get something for nothing.

The idea of getting something for nothing leads me back to where I began, with the Constitution. The takings clause of the Fifth Amendment has been treated by most government officials and all environmentalists as an unfortunate obstacle to the pursuit of the public interest. If I believed it to be an obstacle to the promotion of the general welfare I would suggest that we remove it rather than continually seeking new and creative ways to dodge it. We can recognize an expansive police power or take the more modern approach of excusing our infringements of private rights on the basis of the public trust doctrine, but it would be much simpler to amend the Constitution and be done with the subterfuge. However, I believe either course of action would be a mistake. The Fifth Amendment, with some help from the Fourteenth, is the constitutional foothold of an approach to resource allocation which serves the public interest by serving the self-defined interests of individuals. The very existence of the requirement of just compensation is evidence that those who drafted the amendment believed there would be occasions for public intervention. But they also believed that the public must pay for what it takes. When you get something for nothing, you sacrifice the ability to assess the relative importance of the values that guide human decision.

It is not necessarily a bad thing that forest lands, even those of exceptional timber producing quality, are being converted to urban uses. Central Park in New York City may be an unusually good place to grow hardwoods for furniture manufacture, but few would be in favor of turning it into a tree farm. The land on which Seattle is located was forested at one time, but I have no doubt that society is better off today without the trees. The best use of land depends on the values of the people who might benefit from its use and on the availability of alternative lands on which to pursue those benefits. It would be nice to have all of the forests and the city too, but we have to make choices in a world of scarce resources. Many of those choices have been, and should continue to be, made by individuals acting on the value information they derive from the marketplace. Public decisions are also required, but we must be clear about why they are required and chary of placing too much reliance on government to solve our problems. All to often we begin with the purpose of regulating to correct for market failures and end up in a morass of regulations designed to correct for government failures.

To the extent that we rely on government decisions to allocate urban forest resources, our experience with other land use regulation can be instructive. The institutional mechanisms available include public acquisition, regulation, subsidization, tax law manipulation, public trust, transferable development rights, and other approaches yet to be conceived. Although each is different in its legal definition, and each requires a specialized expertise to be implemented, all of these mechanisms have a common purpose: to substitute some degree of public decision-making authority for private decisions about land use. The difference between public acquisition and a system of transferable development rights is only one of degree. In the former case, all decisions about the acquired lands are to be made by the government. In the latter case, the government decides how much total land in an area will be devoted to development and leaves it to private parties to decide on which permitted lands that development will occur. The fundamental value choice is the same in every public decision, no matter what the preferred mechanism of implementation.

The justifiability of public choices will largely be a function of the decision-making process employed by the government, not the mechanisms used for implementing those decisions. In the American system, the decision-making process is a reflection of the legislative, judicial, and administrative procedures we employ and the level of government to which particular decisions are assigned. Although our land use processes, like most other areas of government decision making, are burdened with mechanisms designed to assure public participation, the central difficulty is always assuring that the public decision is in the public interest. Weighing the expressions of concern, distress, and delight which fill every land use hearing record is a task beyond our most sophisticated empirical techniques. It remains a value choice for those who must ultimately make it. Given this problem of aggregating the values of society's members, local government is more often than not the level of government most likely to reach a defensible decision. However, the conditions that justify public intervention in the first place will also inform our decision about which level of the federal system will best achieve the purposes of public land use decision making (Huffman 1982).

My introductory comments warned that upon fulfilling my technical tasks, I would engage in some philosophizing. Although my musings are both general and abstract, I believe they are relevant and important to the more mundane issues to be addressed in this symposium. The situation at the urban/forest interface has only recently achieved the status of a problem for public debate. Although the previously existing laws preclude any prospect for a fresh start on this particular resource allocation issue, there is still time for reflection on values, and opportunity for assessment of options, before falling in line with the land use regulation approaches of the last few decades. My recommendation is for a lot less law—and with much more clearly defined objectives. But my expectation is that the lawyers, whom we now mass produce, will have more than enough to do in the future.

REFERENCES

Bosselman, F. P. 1981. Does the Lafayette case bring zoning under the antitrust laws? In: E. Netter, ed., Land use law: Issues for the eighties, pp. 50-52. Planners Press, American Planning Association, Washington, D.C.

Huffman, J. 1982. Governing America's resources: Federalism in the 1980's. Environmental Law 12(1):1-74.

Huffman, J., and R. Plantico. 1979. Toward a theory of land use planning: Lessons from Oregon. Land and Water Law Review 14(1):1-74.

James, L. M. 1974. Why forestry professionals should be involved in forest land use planning. In: Foresters in land use planning, pp. 28-33. Proceedings, 1973 National Convention. Society of American Foresters, Washington, D.C.

Klein, S. B. 1980. New state forest practice laws. National Conference of State Legislators, Washington, D.C.

Kusler, J. A. 1980. Regulating sensitive lands. Ballinger Press, Cambridge, Massachusetts.

Land Conservation and Development Commission [LCDC]. 1978. Statewide planning goals and guidelines. Salem, Oregon.

Landmark, T. 1978. Regulation of private logging in California. Ecology Law Quarterly 5(1):139-188.

Liberty, R. L. 1981. Forestland preservation. Harvard Environmental Law Review 5(1):153-183.

Mandelker, D. R. 1976. Environmental and land controls legislation. Bobbs-Merrill, Indianapolis, Indiana.

Merriam, D. H. 1981. Constitutional issues in land use law: Equal protection and the First Amendment. In: E. Netter, ed., Land use law: Issues for the eighties, pp. 113-168.

Netter, E. 1981. Antitrust laws and land use regulation. In: E. Netter, ed., Land use law: Issues for the eighties, pp. 37-49.

Smith, R. M. 1981. Due process: The elements of fair play. In: E. Netter, ed., Land use law: Issues for the eighties, pp. 171-187.

6

THE LEGAL, POLITICAL, AND ADMINISTRATIVE FRAMEWORK FOR THE URBAN/FOREST INTERFACE IN WASHINGTON STATE

Daniel Syrdal

In discussing this subject, because my experience is largely confined to Washington State, I intend to expand on what James Huffman has said by using the Washington experience as an example of how this framework affects the existence and resolution of problems at the urban/forest interface.

As Huffman has noted, land use conversion is not the only, nor necessarily the most important, issue raised by the urban/forest interface. Because of this, and the greater historical experience in Washington with the forest management issues, my discussion will center primarily on management rather than land conversion issues. However, most of the legal and political considerations are equally applicable to both issues, and, as the land conversion pressures at the interface increase, I would expect much of the historical context for management issues to become directly relevant to the conversion issues.

As a result of my personal experience and value system, my perspective is that of an 'environmentalist.' While I certainly believe this to be a valid perspective from which to view the urban/forest interface, it is quite different from that of the typical forest landowner.

This difference can be easily illustrated by looking at the brochure for this symposium. The illustration on the brochure shows a city gradually merging, through various stages, with the surrounding forests and mountains. When a hardcore environmentalist and recreationalist like me looks at this drawing, he may focus on what isn't there more than what is. For example, the stream at the base of the mountain is not shown to be clearcut on both sides with logging debris clogging its channel and blocking fish migration. The logging road at midslope is on a natural

bench rather than cut into a steep and unstable slope, with the resulting soil debris flowing down to the stream below. There is no lower elevation lake with *all* of the trees surrounding it harvested in spite of its heavy use by recreational fishermen. The campers with their tent are limited to those very high slopes nearly at timberline above the logging operations. There are no fish in the lake or stream. The forest landowner's perspective, on the other hand, might be more attuned to the problems associated with the cabin alongside the lake, where some of the best forest land is being reserved for recreational use rather than timber production. He would be very concerned, as some of 'us' might be, about those uses spreading upward to interfere with the harvest of timber off the slopes—that is, unless he also had a subsidiary land development company which could make a large profit from the sale of such lands. He might be thinking that it was okay for that tent to be there for now, but it would sure have to be out of there when his logging truck got up to that next bench. He certainly would be concerned that those other users might affect his ability to harvest all those mature trees.

As is frequently noted, the urban/forest interface is not strictly an area of geographical boundaries, but is also a political interface of differing value systems. The oft-cited article on the urban/forest interface by Henry J. Vaux recognizes this fact: 'If we are to approach the problem of the urban/forest interface professionally, we have to avoid the easy fallacy of assuming that the problem is somehow the fault of politicians, environmentalists, or lumbermen, and face it for what it is: a political conflict over differing value perceptions.'

While, as pointed out by Robert Lee in his chapter, the acceptance of such a definition means that almost all forest policy issues are also interface issues, this does not mean that the physical proximity of urban and forest lands is not a prime contributor to forest policy issues or a definite subset within such issues. Without the geographic proximity of urban lands to forested areas, many of the forest policy issues being faced today would not exist. Fortunately for the recreationalists, and perhaps unfortunately for the forest landowner, modern transportation systems have drastically altered the meaning of 'proximity' in this context. It was not long ago that our present forest lands were largely 'out of sight, out of mind.' Not only was it difficult to reach many of these lands for recreational uses, but there were far fewer people attempting to do so. Furthermore, there was a much greater reservoir of 'unspoiled' lands that could serve as a relief valve for political pressures that would otherwise result from bad management or conversion of forest lands to other uses. More recently, however, increased population, better transportation systems, more leisure time, and money to spend have led to an expanding awareness by urban citizens of the aesthetic and recreational values of forest lands as well as the impacts that forestry activities can have on public resources such as fisheries. This, in turn, has produced a dramatically increasing demand for the use or protection of forest land and resources for purposes other than timber production. Thus the inexorable increase in contact between urban dwellers and the forest environment has led to a conflict in perspectives and value systems over the use of forested lands.

Geographic proximity, however, still plays an important role in the intensity, if not the existence, of problems of the urban/forest interface. In general, the closer one is to an urban center, the more public awareness there is. When one's everyday living is affected by forest land uses, 'back yard environmentalism' tends to intensify the level of apparent concern. Thus, I believe that unless there are radical changes in the way the problems of the urban/forest interface are addressed by the timber industry, we can look forward to greater conflict as the interface becomes increasingly a geographic one.

The Washington experience provides a good case history of how the legal, political, and administrative arenas have been utilized to attempt solutions to the perceived problems of the urban/forest interface. Prior to the 1970s, forest practices in the state of Washington had essentially no regulatory controls. Most forest practices were still conducted off 'in them thar hills' somewhere, with very little public scrutiny. There was little knowledge of the overall cumulative impacts such practices were having on state resources. The timber industry was operated with the typical 'cowboy mentality' of an industry developing a new frontier where external costs could be ignored and private property rights reigned supreme. But as it came to know more about forestry operations, the public became increasingly disenchanted with the external costs imposed on society by the forest industry.

In the state of Washington, as elsewhere, the 1960s marked the beginning of a 'spaceship mentality,' with its recognition of the interdependencies of various systems within an entire ecosystem. People became concerned about impacts to public resources such as air, land, water, wildlife, and fisheries. The public began to accept the concept of the state's obligation to protect such resources even if such protection interfered somewhat with the unfettered use of private property. Urban populations especially came to accept what was seen as necessary encroachment on private property rights for the good of the whole.

This developing difference in perspective and values between forest landowners and Washington's urban population can be seen in the issues over which conflict arose. For example, to the forest landowner and manager, old-growth forest may represent decay, insect damage, a nonvital, overmature forest in need of replacement with vigorous new growth providing a much better yield per acre per year. To the environmentalist or recreational user, however, this forest may provide an aesthetic experience with an open floor for hiking, habitat for cavity nesting species, shade for streams, a source of large organic debris for fishery productivity, a source of soil-building materials, and the provision of soil-holding capacity to minimize sedimentation. Likewise, clearcutting is seen from widely different perspectives. The timber industry largely sees it as the most efficient harvesting method, which also allows for better and quicker reforestation. A recreationalist might see it as a system that maximizes the aesthetic blight of forestry while also threatening fisheries by increasing stream sedimentation, biological oxygen demand, and temperature. By the early 1970s, the political pressure to protect public resources from the effects of development reached the point where various 'solutions' became politically feasible. As Huffman has stated, the federal and state constitutions provide little effective limitation on the power of state and local governments to regulate the forest indsutry, and it had become apparent to 'environmentalists' that regulatory mechanisms were necessary to address their concerns. In the early 1970s, the Washington State Environmental Policy Act (SEPA) and the Washington State Shoreline Management Act (SMA) were enacted. SEPA required, among other things, that before permitting any major action with a potential for a significant adverse impact on the quality of the environment, the government decision maker had to prepare a detailed environmental impact statement (EIS) setting forth the proposal's impacts, alternatives, and possible mitigating measures. The SMA, supported in one or the other of its alternate forms by a large majority of Washington's voters in an initiative election, set forth standards and procedures forming a comprehensive plan for development of the state's shorelines, giving counties the primary enforcement responsiblity.

Unfortunately, these new state laws were largely considered by the timber industry to be inapplicable to its operations. The industry took the position that its operations did not constitute the kind of 'substantial development' subject to the permit requirements of the SMA and that there was no 'major action' to trigger the EIS requirements of SEPA. Those wishing to gain additional control over the impacts of the timber industry on public resources were thus placed in a position of having to resort to the courts or seek new legislation. They did both.

The next chapter of the developing saga came in 1974, when an unlikely coalition of the timber industry and various environmental groups combined efforts to enact the Washington State Forest Practices Act (FPA), supposedly designed to regulate forest practices as they affected water quality. Industry support for this act resulted from the enactment in 1972 of amendments to the Federal Water Pollution Control Act. These amendments (Section 208) were seen as a direct threat of federal regulation of nonpoint pollution sources and provided a tremendous incentive for the industry to support state legislation to head off such federal control. Similar forest practices acts were passed throughout the West Coast states for this reason. Environmentalists, of course, hoped that direct regulation of the timber industry would finally serve to reduce many of its impacts on public resources.

The FPA could have gone a long way in resolving the worsening conflict between the forest industry and various environmental and recreational concerns. The act contains legislative direction to regulate for the protection of forest soils, fisheries, wildlife, water quality and quantity, air quality, recreation, and scenic beauty in addition to the maintenance of a viable forest products industry. However, regulations to implement the act were to be developed by the Forest Practices Board, a board which was dominated by timber interests. In addition, enforcement authority was given to the Department of Natural Resources (DNR), itself a major forest landowner, and, contrary to the statement by Huffman, the Departments of Fisheries and Game were given no regulatory authority under this act and the Department of Ecology's authority was indirect and severely limited. Thus a possible consensus that might have resolved much of the controversy at the urban/forest interface was not achieved in light of the domination of the process by industry and its failure to make significant concessions in the regulations to the value systems of other interested parties.

The first regulations drawn by the Forest Practices Board to implement the FPA clearly demonstrated the control of the industry over this process. The initial draft of the regulations was essentially made up of proposals by the Washington Forest Protection Association. This initial draft was then substantially weakened through successive drafts to a point where many responsible industry personnel admitted that the regulations were less stringent than those practices already conducted by much of the industry. Environmentally oriented regulatory provisions suggested by the Departments of Fisheries and Game, various urban counties of the state, and environmental and recreational interest groups were routinely rejected, often without consideration or discussion. The balancing between industry and environmental concerns contemplated by the legislature simply didn't occur.

In spite of this, a segment of the forest industry in the state of Washington remained adamantly opposed to any regulation at all. Many of the smaller contract loggers were confused by the regulatory process and simply did not feel that their job should be made more difficult by regulations they believed limited their rights as property owners. This opposition resulted in several hundred logging trucks appearing on the state capitol grounds in 1975, certain public officials being hanged in effigy, and a senate vote to repeal the FPA the very day the loggers appeared. This is the ultimate in political responsiveness.

While the larger concerns within the forest industry did not want to lose the protection from federal regulation provided by the act, they also saw this political impetus as an opportunity to provide the industry relief against possible application of SEPA and SMA to forest practices. Legal actions by various local government units and individuals to require such application were being initiated, and it became apparent to the forest industry that it could no longer ignore these state environmental statutes. Thus in 1975 the industry was able to obtain compromise legislation that maintained the Forest Practices Act but provided even greater representation to the timber industry on the board and severely limited the application of SMA and SEPA to forest practices. These 1975 amendments again allowed the timber industry to use its control over the administrative process of formulating regulations to exempt itself from the requirements of SEPA. The Forest Practices Board was, for example, empowered to designate certain forest practices as Class IV forest practices. This class was to include all forest practices having a potential for a substantial adverse impact on the environment. All other forest practices were to be exempt from the EIS requirements of SEPA. Thus, with its large majority on the board, by the time regulations were adopted to implement these amendments, the forest industry was able to define Class IV practices so narrowly that it became essentially exempt from SEPA. For example, of the first approximately 50,000 forest practice applications, not a single application for private forest practices had fallen within the Class IV designation requiring the preparation of an EIS, even though the legislative history of the 1975 amendments clearly indicated a legislative intent that about 5 percent of all forest practices would fall in this category.

Furthermore, contrary to Huffman's statement, the FPA does not necessarily require environmental impact statements for forest land conversions. The SEPA requirements of the FPA are a separate category under Class IV which has recently been designated by the board as Class IV Special. The act simply doesn't speak to the question of the preparation of EISs for other forest practices related to forest land conversions.

The Forest Practices Board had essentially written Class IV out of existence even though the statute required that all forest activities with a potential substantial impact on the quality of the environment be included within such a class. Obviously this represented a great disparity in perspective and value judgments between many of the environmental and recreational interests that had sought enactment of this act, and supported the 1975 compromise amendments, and the Forest Practices Board.

The success of the timber industry in blunting the statutory efforts of proponents of greater protection for public resources on private forest lands in turn caused those proponents to go to the next stage in the legal arena: court challenges. In 1977, an action (NOEL v. Cole) was brought against the commissioner of public lands as head of DNR. This lawsuit, commonly known as the Classic U case, challenged the propriety of a state timber sale on lands adjoining a rural state park on Whidbey Island. It was contended that the DNR was required to prepare an EIS prior to the timber sale. Until that time, DNR had, consistent with the action of other forest landowners in the state, failed to prepare any environmental impact statements on its timber sales or forest practices. The court in the Classic U case not only held that DNR was required to comply with SEPA by preparing an EIS on that timber sale, but it also held that the Forest Practices Board's definition of Class IV forest practices was void on its face and 'so narrow and restricted as to almost totally thwart the purpose and intent of the legislature.' The court in that case went on to note: 'The court is compelled to comment that a suit such as the instant one was inevitable in view of the overzealous actions of some state agencies in removing forest practices from SEPA considerations. . . . this is not in keeping with either the intent of the legislation, nor this state's policy of endeavoring to balance environmental and non-environmental interests. . . the elimination of almost all environmental considerations is not in keeping with legislative purpose and intent and leads directly to litigation of this kind.'

In a similar piece of litigation, King County, the most populous county in the state, succeeded in having the Washington State Supreme Court declare the 1975 limitations on the application of the SMA to forest practices as having been unconstitutionally adopted and therefore void. The court also rejected in the strongest terms the industry's long-standing contentions regarding the inapplicability of the SMA to its actions.

After the Classic U case, the forest industry and DNR persisted in their attempts to avoid any SEPA compliance with respect to forest practices. In litigation challenging the failure of DNR to require the preparation of an EIS for certain private forest practices, the agency took the position that the Classic U case was only binding on it in Island County and that it therefore could not be the basis for requiring private applicants to comply with SEPA in other counties. Furthermore, the Forest Practices Board had, more than two and a half years after the decision in Classic U, made no attempt either to broaden its definition of Class IV Special to include forest practices deemed by environmental and recreational interests to be especially detrimental or to respond to the decision in any way.

By this time it was apparent that the forest industry and the Forest Practices Board intended to continue to ignore SEPA as long as possible and until the board was under direct legal compulsion to do otherwise. The fact that the timber industry seemed to have a greater impact on the natural environment than all other industries, and these other industries did have to comply with SEPA, made this situation harder to tolerate. This seeming intransigence led various sports fishing and environmental organizations to bring a state declaratory judgment action against the DNR and Forest Practices Board to force compliance with SEPA on a statewide basis. This case, known as the Steelhead Trout case, is currently pending and seeks statewide application for the Classic U decision as to the invalidity of the Class IV regulations and a requirement that the regulations be revised to include all actions with a potential for a substantial impact on the environment, as the FPA provides. For the first time, this lawsuit provides the possibility of direct court intervention in the formulation of forest practice regulations.

After the election of a new public lands commissioner, Brian Boyle, who is also a contributor to this symposium, and the filing of the Steelhead Trout case, the Forest Practices Board finally began a process of reviewing its regulations and considering additions to the Class IV Special designation of forest practices. This review recently culminated in the adoption of certain amendments to the forest practices rules, including amendments as to what forest practices constitute Class IV Special.

While the new regulations, for the first time, provide the possibility that a substantial problem area will be reviewed under SEPA provisions, there are still many substantial impacts that are not dealt with by the new regulations. Compared with the old definition, the major change has been the possibility of SEPA review for forest practices involving the construction of roads, landings, and so forth, on excessively steep slopes or slide-prone areas that are in close proximity to streams that have a potential of being significantly affected by debris flows into fishery habitats or public capital improvements. The regulations don't explain what constitutes an 'excessively steep slope' or how the department will determine, based on the information provided in a forest practice application, whether there are slide-prone areas in sufficient proximity to important fishery habitats, and so on. The Forest Practices Board rejected several other proposed inclusions to Class IV Special, some of which were supported by the DNR and various technical review committees and interest groups.

Even with the new inclusions in Class IV Special, I believe it is clear that the Forest Practices Board is still neglecting its obligation to strike an appropriate balance between economic considerations to the forest industry and the measures necessary to protect public resources. The forest practices involved in the Classic U case, which were required by the court to have an EIS, would still not be subject to an EIS under the new regulations. The regulations still, in spite of the clear statutory mandate to the contrary, totally ignore scenic and recreational values as far as either regulatory protective measures or application of SEPA is concerned. For example, one could still clearcut several thousand acres adjoining Rainier National Park without so much as considering the recreational or scenic impacts and still be in full compliance with the forest practice regulations. The regulations also fail to deal with some of the most pressing issues associated with forest operations—issues that can only be dealt with in the context of SEPA. Such things as the cumulative effects of a large number of forest practice operations in a given watershed, and the offsite and social impacts of a given proposal, cannot be dealt with by a regulatory approach applicable to all cases, but must be considered on a case-by-case basis. This is only possible through the kind of review provided under SEPA and the Shoreline Management Act. Generally applicable regulations under the FPA would have to be unduly restrictive in order to be effective, and thus would run counter to the industry's best interests.

So the basic question remains where all these political and legal machinations are leading us with respect to the urban/forest interface and the statutory and administrative framework in which it will operate. I suggest that the forest industry will have to decide whether it will continue its stonewalling approach or recognize the need to accommodate other value systems. While I am confident that the ultimate result of the conflict will lead to much greater resource protection in the future, I am unable to predict the legal and administrative framework that will get us there.

My willingness to predict the ultimate 'victory' of additional environmental protection in the practice of forestry in Washington results from my firm conviction that the progress of this country away from a frontier approach to its natural resources and toward the spaceship concept cannot be reversed. On both the state and national levels, recent polls have shown that the vast majority of our population supports the maintenance of strict regulatory controls in order to improve environmental quality. For example, in a recent poll conducted in the state of Washington, at a time when both unemployment and the rhetoric opposing regulation have been extremely high, 68 percent of those polled indicated that environmental protection is necessary in spite of the fact that economic growth should be a high priority. Of those polled, 78 percent favored the expansion of our recreation and tourism industries as important state goals. In addition, fishermen of the state are becoming increasingly aware of the direct impacts that the timber industry has on their way of life. The federal courts have declared the Indian fishermen of Washington to have special treaty rights to protection of the environment—rights that can only serve to increase the pressure on the timber industry to improve its practices with respect to water quality. Non-Indian recreational and commercial fishermen have also become increasingly aware of the impacts of the timber industry on the fishery resource. A telling sign of the political situation the industry will face in the future was seen at the Washington legislative session of 1982. For the first time, warring Indian and non-Indian fishing factions worked together to defeat industry-supported legislation that would have weakened environmental protection of the state's fisheries. These fishermen are not willing to continue incurring the external costs of the forest industry's failure to conduct forest practices in a way that is protective of the fishery resource.

Because of its perceived unwillingness to address its impacts on other forest users, the timber industry in Washington is, I believe, creating a far more powerful opposing coalition of various fishing, environmental, and recreational interests. If this is true, the major forums for future resolution of the issues of the urban/forest interface will undoubtedly be the courts and the legislature. As more and more people live close to lands used for commercial forestry, there will undoubtedly be an increase in the number of direct damage actions brought against the industry. As the fishery resource becomes scarcer, the timber industry's impact thereon will be more closely scrutinized and litigation concerning damages can be expected. If necessary, litigation and legislation seeking a more adequate control of poor forest practices will be pursued, and will be successful.

There are other possible avenues for resolving the issues pertinent to the urban/forest interface. Because of the inherent inflexibility of the legislature and the courts, I don't believe resorting to these forums is the best way to resolve the conflicts. For example, the various interest groups representing different views on the use of forest lands could more probably reach an acceptable resolution of many of the issues regarding appropriate regulation than could the courts or the legislature. As far as management issues are concerned, Washington already has the legal framework to provide an adequate regulatory format to protect the public resources against unnecessary damage by the timber industry. All that is required is agreement on what regulatory provisions are necessary and a willingness to take reasonable measures to avoid unnecessary adverse impacts on a case-by-case basis. However, while there have been some hopeful signs in recent actions by the DNR and private timber interests, a negotiated resolution is certainly not possible if the industry takes the position that a meaningful regulatory format is not necessary to solve the problem.

The controversy about the applicability of EIS requirements to forest practices highlights another very real problem in the way the legal-administrative framework has worked in the state of Washington. Because, in a legal or administrative framework, the opposing sides are forced to be more dogmatic than pragmatic, and to simplify issues rather than admitting their complexities, the antagonists seem not to realize what each other really wants. For example, time and again I have heard the strongly felt belief expressed by forest landowners that the environmentalists intend to see that an EIS is required for every forest practice undertaken in the state. This is certainly not the case. At the same time, many forest landowners would probably agree to substantial modifications of their practices when necessary to provide protection to public resources. This was, however, far from the message received by environmentalists, probably because the industry felt it must take a unified stand to minimize the regulatory intrusion in its operations. Unless a different forum is provided to allow for better communication and negotiation, these misperceptions will probably continue and will necessarily result in the legislature or the courts being the final arbiter of conflicts at the urban/forest interface. This, I think, will clearly work against the best interests of the forest industry.

7

LAND USE PLANNING AT THE URBAN/FOREST INTERFACE

Glenn J. Amster

The complexity of the laws affecting the allocation of resources at the urban fringe is often overwhelming. This presents a dilemma to the property owner contemplating the potential use of his land, the governmental agency attempting to decide on an appropriate use of that land, and the lawyer asked to advise his client on a solution. The public usually blames government for creating so many legal complexities and subsequently blames lawyers for having become a necessary part of the process. In fact, as James Huffman recognizes in his thoughtful chapter, the bulk of the law is the result of political pressure and the increasing complexity of the law is a result of the increasing fractionalization of the public interest. The diversity of public views causes the government to prescribe a burdensome process—after all, everyone's views must be considered, if not accommodated—which inevitably leads to an inefficient allocation of resources.

The traditional assumption is that timber production and other forest uses are incompatible with urbanization. In fact, as we are becoming increasingly aware, urbanization is of prime importance to the forest resource because it represents a market that may be served by the forest and its products. Without this market demand for forest products, there is greater pressure to convert our forests, perhaps at a more rapid rate than would otherwise occur, to nonforest uses. The private sector, apparently able to recognize and resolve its own differences, has long ago discarded this traditional assumption. Unfortunately, it is an assumption that seems to continue to plague the land use planning process.

When preparations are being made for a comprehensive plan, it is not unusual to find diverse interest groups, environmentalists, landowners, home builders, and government staff agreeing on the basic objectives of a long-range plan. They all probably agree that the plan should guide anticipated growth into development patterns that use land efficiently and wisely, provide affordable housing and other development opportunities, encourage productive use of farm and forest lands, and provide recreational and other amenities while maintaining a healthful environment. Most would agree that these goals should be accomplished at reasonable costs and within generally accepted political and legal frameworks, with the rights and aspirations of the public balanced fairly against the rights and obligations of landowners and businesses. But what we all too often see emerging from our planning departments and from community-based planning committees are guidelines and policies stated in the negative—restraints on land use change and restraints, rather than encouragement, of certain development patterns.

In many instances, it has become easier to decide issues of resource allocation by not deciding them. A planning technique that reveals this kind of approach is the concept of the 'growth reserve'—the designation of a boundary that at least for the near future is to mark the outer limits of growth. It does not matter where the line is drawn, and we can disregard for now the issue of how that line may have been determined (though the longest and hardest debate often stems from these issues because this technique represents the ultimate market manipulation—the removal of certain lands from the available supply). Rather, it is the concept itself that is fatally flawed. No one can assume, or designate, an area to be frozen in time or, at the urban fringe, not yet committed to growth.

The 'growth reserve' concept essentially stems from a desire to promote infill, arguably because this development pattern 'maximizes' existing services and utilities and thus is deemed cost effective. An underlying premise is that services and utilities are available or can easily be made available within already developed communities. There are several fallacies to this approach. First, it assumes that local government will commit its fiscal resources to infill areas so as to ensure that adequate services and utilities are provided and can be maintained. Given present economic conditions, this assumption is questionable. Second, notwithstanding the issue of cost, it is often much more difficult to expand existing facilities within already developed areas than to construct new facilities in undeveloped areas. For example, imagine the uproar that would occur in an urban community over an attempt to widen a two-lane road to four lanes, as opposed to developing a four-lane road to begin with, before development occurs.

A planning strategy based principally on an infill concept has another basic flaw: it ignores the benefits to the urban community of skipped-over land. For example, skipped-over land has important open space value to residents in urban and suburban communities. We all can appreciate the aesthetic and social benefits of our urban forests. It also provides an opportunity for development over widely varying periods rather than at the same general time. Not only does this contribute to the vitality of the community, but it seeks to prevent wholesale neighborhood deterioration. To require the filling in of skipped-over land before new areas can be developed thus may have significant social impacts apart from the physical impacts directly attributable to growth. While we may not want to discourage infill, these factors indicate that an overall growth strategy based on infill may be undesirable.

A corollary planning technique is the designation of 'rural communities.' The designation is intended to preserve and protect rural areas from the pressure of growth. The concept of 'rural' communities, however, often reflects a nostalgia for the past rather than an image of existing development patterns and philosophies. For example, rural towns in such plans are generally defined as small clusters of rural-type businesses, such as feed and seed stores, farm equipment distribution centers, and repair and other businesses serving agriculture, mining, and forestry industries. These time-worn definitions fail to recognize the vitality of existing rural areas, their desirability as places to live, and thus their need for economic diversity. Recent surveys, and information gathered from the 1980 census, indicate significant growth in rural towns. Apparently, despite long-distance commutes and high commuting costs, many people are willing—even eager—to live relatively far from urban-suburban communities and existing employment centers.

From the urban/forest interface perspective, the problem with 'growth reserve' and 'rural planning' policies derives from the large-tract land use designations that typically implement these policies. A key assumption in this strategy is that large minimum lot sizes will discourage development and effectively shift residential demand to the urban-suburban areas. But there are other effects that are likely to occur. First, because of higher land costs in developed areas, discouraging growth in rural areas is likely to lead to less new construction. This will have an adverse effect on the economy, the availability of jobs, and the cost of housing. More important, some growth inevitably will be displaced to the reserve or rural areas. In other words, land in these rural areas will be developed in a suburban pattern and thus will be unavailable for later development at higher densities. As a result of this shortsightedness, major infrastructure improvements that may be necessary in the future are not planned for, and planning staffs and local governments are not able to address the issues of growth once development becomes imminent.

Another function of typical 'rural' use designations is a restriction on the development of industries in rural areas to those that are dependent on resources in the area, such as agriculture, mining, and forestry. Forcing rural areas to depend on resource-based industries ignores the notoriously cyclical nature of these industries and their susceptibility to sharp fluctuations in the economy. Preventing rural areas from diversifying their employment base may lead to economic stagnation. Thus when the economic cycle swings up, and the resource-based industries regain economic vitality, they will have to look elsewhere for employees; they will find that the unemployed rural population has left, and they will have a difficult time attracting new employees because of the scarcity of housing and job opportunities for their families.

Aside from the potential effect on housing starts and job opportunities, rural area land use policies that substantially limit development in the foreseeable future may pose a serious threat to the productivity of forest resources. It seems that management of forest lands is not threatened by traditional subdivisions (half-acre lots and less) so much as by scattered development on larger tracts. In the first place, large-lot development consumes more land. Since the demand for housing would remain fairly constant, large-lot zoning means that more land will be used to meet this demand. Second, a few widely scattered 'neighbors' are more troublesome for the forest manager than larger numbers of people clustered in a few small-lot subdivisions. When the industry seeks to build roads, harvest timber, burn slash, apply chemicals, or take the other steps essential to modern forestry, the number of neighbors itself is not a problem so much as the number of points within the commercial forest area where there is some development, even a single home.

Dealing with such planning policies and techniques requires innovation and the discarding of traditional values and assumptions. Although some would suggest that we stop wasting the taxpayers' dollars on planners because the results may not be significantly different without them, there are considerable benefits to be derived from proper land use planning. Our efforts should be directed at developing long-range plans that recognize the realities of the development world, and creating the type of private-public partnership that will be so essential in the future.

If a land use plan is to be based at least in part on an infill policy, government agencies must recognize that there will be resistance by existing residents to infill in their communities, and that there are valid reasons for this resistance. To the extent that infill is desired, local governments must own up to their own limitations as far as the private-public partnership is concerned, and accept responsibility for the growing pains that the affected communities will experience. Furthermore, given the widespread deficiencies in existing infill areas and the limited financial resources, local governments must offer something positive to encourage development in these areas. The private sector cannot simply be told to move into these areas, pay higher land costs, and make up for whatever deficiencies there may be in the way of public facilities.

Incentives could include fast-tract development approvals, density bonuses for services and utilities that require little expansion, flexibility to allow innovative solutions for capacity limitations (e.g., high-occupancy vehicle and housing alternatives), and predictability in development approvals. An aggressive infill policy may also require that we lower our expectations with respect to levels of public service, particularly in transportation modes, since the promotion of infill often is limited by policies requiring 'adequate' capacity of existing services.

If in fact we want the rural area to be preserved for rural uses, planners and government agencies must recognize the true nature of those uses. They must acknowledge the 'undesirable' aspects of resource industries and their effects on surrounding development. If, for example, our interest is in preserving a viable forest industry, protection of that industry may not permit 5-acre (or larger) residential developments on the perimeter of forest land, but may instead require specific concentrations of density, to keep scattered development from jeopardizing timber operations. High-density development in certain areas, governed by specific design and performance standards assuring appropriate quality control, is probably more desirable.

As certain forest land is set aside for timber operations, there must be acknowledgment that it will be logged periodically and its aesthetic and open space value diminished. Development of adjacent lands need not conflict with commercial forest management. Persons who develop in and near commercial forest lands, and persons who occupy and use these developments, can plan their activities with the understanding that forest management activities will occur periodically on the nearby forest. At the same time, areas within the densely developed portions of the rural area should be set aside to satisfy recreational and open space needs.

Land use plans should designate new urban-suburban communities and encourage the cost-effective and energy-efficient option of large tract development. Local governments must be prepared to make a genuine commitment to the development of these communities, since they will be the only areas—whether new industrial areas or residential communities—able to accommodate growth in an appropriate manner. In addition, planning policies and regulations should encourage the development of a diversity of uses in rural areas.

Of fundamental importance in developing new communities is the understanding that these developments are often 10 to 200 year propositions. If the private sector is required to undertake its own long-range planning in order to provide 'desirable' development, then surely local governments should be willing to look to the future. Thus in planning for large development areas, or designating new urban-suburban communities, plans must include criteria that enable us to anticipate needed routes for major utility and road corridors. In addition, the process for reviewing such proposals must be delineated and local government must enlist the support of the private sector in undertaking the type of long-range planning these developments require. A desired population-employment balance can be obtained in the outlying areas as well as in proximity to existing growth centers. Local government must recognize, and must undertake to educate the public, that different land uses, including light manufacturing, forest management, and office parks, can be made compatible with residential uses.

Streamlining the process also requires reversing the relatively recent use of environmental protection laws as an obstacle to development. Not only does such use undermine the legitimate purposes of these laws and regulations, but it also results in higher costs for the public and private sectors, as well as the consumer. It has become acceptable practice for opponents of a project to challenge it at least through the administrative processes, and more often than not through the courts, with the hope that the developer will lose interest or that the project will lose its economic viability.

The environmental review process should be revised substantially to eliminate or mitigate current abuses. The threshold for an environmental impact statement should be raised considerably and exemptions should be provided to a broader range of uses. We may want to consider, for example, exempting all residential subdivisions of less than a certain size from the EIS requirement if such development is consistent with the zoning. The EIS format should be modified to require discussion of only those issues of significance in a particular situation. In a strictly legal sense, this is the standard now, but the ability to challenge an impact statement and the subjective nature of the 'adequacy' test require that the EIS describe every detail, regardless of its significance, particularly in controversial situations. Another solution may be to limit standing to challenge the adequacy of an EIS to those with a direct interest—those who are directly affected by a proposal—or to those who have participated in the process. It is frustrating when someone who has not participated in a public hearing or offered constructive comments during the circulation of a draft EIS challenges the final EIS because his concerns have not been addressed. Finally, we may even want to consider eliminating all challenges to the adequacy of an impact statement, leaving a challenge to the substantive decision itself as the testing grounds for determining whether environmental issues have been properly considered.

The situation at the urban/forest interface only now is beginning to demand the attention it deserves. We can expect a thorough and inevitably prolonged public debate on the values and resources at stake. Before the debate subsides, the decision makers will be asked to allocate some of these resources and choose between conflicting values. Whether they are up to the task will depend in large part on the ability of the public and private sectors to adapt and adjust the existing legal processes to coincide with changes in the economic and social environment. Planning precepts suited to the needs of decades past, therefore, cannot be relied upon to meet the needs of the developing society in the years to come.

8

PUBLIC LANDS: PAST, PRESENT, AND TOMORROW

Brian Boyle

I considered discussing the state of Washington as a regulator of forest practices and the process of public decision making. However, I believe it is more important to discuss the state as owner, rather than regulator, of forest lands. I want to comment briefly on Professor Huffman's chapter, and to outline the background and history of public landownership in the state. Then I would like to discuss the state's legal responsibilities for managing public lands, and finally discuss how urbanization affects state land management.

Professor Huffman asserts that motivations for land use regulation have 'little to do with efficient resource allocation and a lot to do with wealth redistribution.' I certainly would not dispute that economic gain and quick profits have played a large role in our nation's approach to land use. Land and its potential wealth motivated the English trading companies to organize themselves and sell stock to establish American colonies.

But I would like to use the same liberty with Professor Huffman's discussion as he proposes for the symposium title. He suggests including the word 'choices.' I think his proposal has merit. This urban/forest interface issue is a question of choices in our society—of how our lands are to be used. But I want to expand his legal and economic framework for land use decisions to include social values. As Henry Vaux has written, the urban/forest interface—or interaction—is between 'different political forces, each of which has a somewhat different set of perceptions about the value of forests to society' (Vaux 1982). These values may not be calculated in any meaningful monetary fashion. Rather, they may merely be something intrinsically desirable. We can see in our nation's early history that while many colonists wanted the free land, others were motivated to migrate to find a place to practice their religion. These actions were not strictly economic decisions. They represent land use decisions based on social values.

These social values have been intertwined throughout our nation's land policy decisions. Land promises or preferences were given to veterans of the French and Indian War as well as to those of World War II. Several measures were passed in the 19th century to set aside public lands for cemeteries, churches, or monasteries. Cities and counties received land for jails and buildings, and shortly before the Civil War, Congress voted to give land in northern Illinois to Polish exiles from Russian tyranny (Dick 1970).

Throughout the 19th century, there were battles over homesteads, migration policy, and public land sales. This again intermingled social, political, and economic issues. Should landless settlers receive free land in the West while the government gave up land revenue and encouraged people to leave the eastern seaboard? It certainly can be concluded that the Homestead Act did not fulfill its original intent to give every poor person land to have a home. But the slogan of 'free homes for the millions' was a magnet that drew million to the unsettled West, and the results after 120 years have been spectacular.

Starting with a land grant to a canal project in 1827, the federal government began to give land to encourage the construction of transportation projects. This ultimately led to grants totaling 155 million acres to railroads to build new lines to the undeveloped West.

Besides underwriting a national transportation system, land grants were used to achieve an important social goal—the creation of a public, nonsectarian educational system. The nation's land survey system, outlined in the land ordinances of 1785 and 1787, set aside section 16 of every new township for public schools. Later as many as three other sections of every township were also allocated to schools. The purpose was to encourage 'the means of education' in the new states (Carstensen 1963).

Another use of land to establish a system of public education evolved from a 1787 contract between the federal government and the Ohio Company. The contract required two townships to be used for a university. This was the start of federal land grants for universities, culminating in the Morrill Act of 1861 which established public land endowments for colleges of agriculture and mechanic arts across the country.

There is no doubt about widespread fraud and misappropriation of these school and university lands. But these land grants evolved out of society's desire for public education, not just profits and wealth redistribution. Washington State's public landholdings have largely resulted from the social values embodied in the federal land dowries for new states entering the Union. This federal endowment by 1889 had grown to include land not only for schools and universities but also for an agricultural college, a scientific school, construction and maintenance of public buildings at the state capitol, state normal schools, and state charitable, penal, and reformatory institutions.

Washington's total land grant from the federal government was about 3.1 million acres—or about 6 percent of the state's land mass.

What is remarkable about Washington State in contrast to its neighbors is that it has attempted to carry through the social objectives of the land grants for nearly one hundred years. While sizable portions of the state's trust lands have been sold, polls, editorial comment, and legislative actions would reflect firm agreement with one 1889 commentary about the Washington constitution: Washington's public lands, the commentary said, will be valuable for agriculture or timber and 'they ought not to be frittered away' (Hill 1889).

A consistent strain throughout Washington State government has been a wariness of special-interest influence on government. As a result, the state constitution continues to direct that public lands be held in trust 'for all the people' and that trust lands only be disposed of when fair market value is received (Washington State Constitution, Art. XVI, sec. 1).

The state expanded the public landholdings when cutover forest lands became tax delinquent in the 1930s. About 80,000 acres were purchased by the state for about 50 cents an acre. Another 540,000 acres of forest lands were deeded to the state for support of counties.

So today, when one subtracts the sold land, Washington's public landholdings comprise about 2.4 million acres of federal trust land and another 622,000 acres of County Forest Board Lands. These state lands include 1.8 million acres of commercial forest land, about 10 percent of the total commercial forest acreage in the state.

What does this public land base mean today for Washington citizens? First, it produces income for schools, universities, counties, and state institutions. This biennium the state expects to generate about $350 million dollars from trust land timber sales and leases. This income is a significant source of construction financing for public schools and universities. It also supports resource management work such as reforestation and fertilization. Second, state lands fulfill other social objectives besides production of income for vital institutions. One of these goals is to operate state lands on a multiple use basis, and it has led to the establishment of 565,000 acres of multiple use areas on state forest lands. State lands also have been included in land management and environmental laws such as the Washington State Forest Practices Act and the State Environmental Policy Act, and federal laws such as the Clean Water Act and the Endangered Species Act. These objectives again reflect society's *choice* of values involving the use of natural resources.

It must be conceded that these different social objectives sometimes conflict. But as Vaux suggests, the fact that society has placed different, and sometimes conflicting, values on the forest is not a new trend. Differences between political groups, he states, lie 'at the root of every major forest policy debate at least back to the Magna Carta' (Vaux 1982).

The debate about the social values of the forest has focused in recent years on land affected by a new shift of population from urban metropolitan areas to rural nonmetropolitan areas. J. C. Doherty, in another recent article, calls this population movement 'possibly the most influential single trend—social, economic, or political—on the national scene in the 70's' (Doherty 1981).

If you happen to drive east across Lake Washington into central and eastern King County, you will see abundant examples of what Doherty calls the 'countryfied city.' Your observations will be borne out by a recent statistical analysis by Richard Morrill. He found that between 1970 and 1980 the core cities of southern Puget Sound lost 50,000 people, or 5 percent of their population. The suburbs gained 230,000 (a 33 percent increase) and the areas beyond the suburbs, but within commuting distances of major cities, recorded an increase of 170,000 (55 percent). Such attractive recreational areas as Washington's San Juan Islands saw their populations double in ten years (Morrill 1981).

In Morrill's survey, a majority of the people he interviewed said they came to the Pacific Northwest for 'environmental and social reasons—not for specific jobs.' These newcomers especially wanted settings that were suitable for recreation, second homes, or retirement life. Noting that this Pacific Northwest phenomenon reflected a national trend, Morrill concluded that 'Americans evidently have not yet given up their search for "our own"—for that pastoral sense of community and shared values that the urban America of recent decades seems hard pressed to provide.' Thus it is more than wealth redistribution that motivates this population shift and this intense conflict of urbanization on the forest frontier.

Your tour of suburban and rural King County, however, might not show how much this urbanization conflict affects state trust lands in Washington. First, there is the statistical picture. Approximately 1.4 million acres of state-owned commercial forest land are in western Washington. Population growth in the last ten years exceeded the state average in more than two-thirds of the western Washington counties. More important, these counties whose populations are growing faster than the state average contain more than 900,000 acres of state-owned commercial forest land.

These lands are high quality sites since they have good soils, ample rainfall, and long growing seasons. They must be kept in our forest base to allow these timberlands to produce income for education and to continue to provide a source of timber supply to many mills and companies.

Next, we must consider operational or forest management implications. For example, one of these fast-growing western Washington counties has zoned all state lands as conservancy, ruling out any timber cutting without a special use permit. Another dilemma is the forest fire problems created by the growth of suburbs near forest areas. A third example is a recent experience in which we set aside plans for a proposed sewage sludge application on state land on the Tahuya Peninsula southwest of Bremerton after area residents, including several persons who had moved from California to establish a new life in rural Washington, protested vehemently.

In the first instance, we are confronted with a state versus local primacy problem. In a fashion similar to Professor Huffman's arguments for federal primacy, we attempt to resist asserting state exemption from local land use ordinances. However, local government must accept the statewide interest at stake, and not attempt to spot zone state forest land and thereby diminish economic returns to school districts statewide.

In the second case, areas—particularly in eastern Washington—that have traditionally been swept by wildfire are now sites of rural subdivisions. Our state forestry agency is responsible for forest fire control, and frequently finds itself unable to provide urban-type fire assistance to homes caught up in rural wildfire.

In the third case, immense energy outlays for sewage sludge incineration can be saved by dispersing the sludge on forest lands. The resultant forest growth is impressive, but may never pass the test of a semirural constituency.

Clearly then, the problem is here already on state lands in Washington and likely to grow in severity. It has already led to significant changes in state trust land management. These changes include a different management direction for lands threatened with urban development, including a need for more public contact and education, and new site specific planning.

The first change prompted by the urban/forest conflict has been a reevaluation of the management of about 120,000 acres of state trust land that lie in or near urban areas. Some lands are already within municipal boundaries. Other parcels face the strong likelihood of development pressures in the next forty years. There is a 640-acre school trust section east of Lake Sammamish that has a residential subdivision about a half mile away. We no longer consider this or the rest of the 120,000 acres to be suitable for traditional agricultural or forest management. Instead, we have developed a transitional lands program and must consider either more innovative forest management or eventual conversion to urban uses.

But this means we must work closer with affected interest groups and the public to reach agreements on forest land issues. For the Department of Natural Resources, this has included the spelling out of our herbicide plans for a group of concerned northwest Washington residents, agreeing to consult the city of Everett prior to timber sales in its watershed, and finding a way to save a group of old-growth trees beloved by Whidbey Island residents.

I have found that these groups recognize our trust responsibilities when we explain our management plans and in some cases modify those plans to meet their objections. The result can be a reasonable compromise.

In the case of the old-growth trees on Whidbey Island, an early proposal was to cut them down and replant the area. Whidbey Island residents objected to these plans and took the matter to court. The result was not only a delay in the timber sale but also a court decision that caused the state to halt all timber sales for several months. The cost to the state trust beneficiaries exceeded $4 million. In addition, and more important, the judge's decision forced the state to rewrite the Forest Practices Act regulations—a process that took about three years.

We have now agreed to seek a way to preserve the old-growth trees while also compensating the trust beneficiary—in this case, the University of Washington. It appears that we will be able to trade this forested site with the State Parks and Recreation Commission for timberland elsewhere.

We must make greater efforts to discuss the forest land management program with the press and civic groups. It is only by such discussion that we can explain both sides of such issues as clearcutting, slash burns, and the cutting of old-growth trees.

We have recently proposed to acquaint Seattle area residents with the forest industry through the creation of the Tiger Mountain State Forest just east of Issaquah. This project is designed to maintain about 13,000 acres of forest land while providing recreation, forestry education, and trust land income. It is an ambitious proposal, but I believe we can be successful.

We are now beginning a new site specific approach to our public timber sales. This will be part of a new forest land management plan, the result of about two years of work. Forest managers will be given increased responsibility to consider a specific timber sale site and decide the shape and manner of the sale. It will give them flexibility to fit public needs into state requirements to maintain the timberland productivity to provide future benefits to education.

What I am urging foresters to do is to face the problems of urbanization and not retreat from them. Our forest management cannot pull back to the mountains. For one thing, much of that mountain timber already has been cut. The lower Puget Sound Basin second growth is at or nearing maturity. So unless urban/forest conflicts are reduced, it will be difficult for Washington to maintain the volume of its timber harvest.

In conclusion, I believe Vaux summarized the approach very well when he concluded that forestry decisions are being made in an environment 'in which political factors are at least equal in importance to technical ones in determining how forestry will be practiced. What it really boils down to is that the forestry profession needs to develop skills for dealing with people to add to the skills used for dealing with trees' (Vaux 1982).

The state of Washington is more limited in its choices than a private forest landowner. Public expectations dictate hunting, fishing, recreation, beauty, and timber resources at the same time. Public perceptions of state lands, or at least local planners' perceptions, are clouded by visions of potential parks. But we do have a powerful argument for forest productivity. The social value of forest management is directly linked to our state's educational system.

That link between natural resource management and education could well be the point that finally relieves public pressure on both private and public forest lands in this state, and results in some compatibility between urban growth, local planning, and forest productivity.

REFERENCES

Carstensen, V. 1963. Introduction. In: V. Carstensen, ed., The public lands: Studies in the history of the public domain. University of Wisconsin Press, Madison.

Dick, E. 1970. The lure of the land: A social history of the public lands from the Articles of Confederation to the New Deal. University of Nebraska Press, Lincoln.

Doherty, J. C. 1981. The countryfied city. American Land Forum 2(Fall):7-8.

Hill, W. L. 1889. Washington: A constitution adapted to the coming state. Typewritten copy of an article printed in the Morning Oregonian, Portland, Oregon 7/4/89.

Morrill, R. 1981. Census 1980—Northwestward ho: Moving in on the last frontier. Pacific Northwest 15 (July-August):18-21.

Vaux, H. J. 1982. Forestry's hotseat: The urban/forest interface. American Forests 88(5):37, 44-46.

9

IMPLICATIONS FOR PRIVATE FOREST MANAGEMENT

Jack Larsen

James Huffman provides an excellent description and analysis of the constitutional, federal, state, and local basis for government actions affecting public and private allocation of urban forest resources. In addition, I found his commentary on what we should do at the urban/forest interface thoughtful and constructive. I cannot disagree with any of his major points. The emphasis of my commentary is therefore on further discussion of what we should do within the current legal, political, and administrative framework as private forest managers.

The forest products industry entered the 1980s with a huge resource base but the worst economic climate most of us can remember. In managing their assets, companies will need to recognize and account for constraints placed on them by government to meet the public's goal of clean air and water, availability of open space, and aesthetically pleasing development of land. These constraints will be especially complex at the urban/forest interface. Whether government can provide a regulatory framework to meet the public's interests, yet encourage wise allocation and investment in forest management, remains to be seen.

General principles that should be considered for such a regulatory framework include:

1. Reliance should be placed on free market principles to allocate resources and permit expression of individual preference.
2. A high degree of predictability and consistency in procedural actions should be sought.
3. Decisions regarding land use should be made at the lowest appropriate level of government.
4. Programs related to the use of land should be designed and administered to affect individuals and firms evenly and be flexible enough to accommodate unexpected conditions.
5. Consideration should be given to a broader range of regulatory devices than zoning or use prohibitions.
6. Local land use planning should be accomplished on a comprehensive basis around specific goals and policies appropriate to accommodate long-term economic growth, and by utilizing readily available information on the geological, physical, and biological characteristics of the land.
7. Government bodies should be required to reach decisions without the undue delays currently being experienced in many procedures.
8. The numbers of permits and approvals currently required should be reduced and better coordinated between agencies.
9. Any identification of areas to be used primarily for private commercial forest crops should be made by local (rather than state or federal) government. And any

such identification process should be kept separate from regulation of forest practices.

10. Regulations should be flexible enough to allow for other compatible land uses in forested areas and for owners to change land use as demand and need for land change.

11. It should be recognized that harvest and regeneration are integral parts of forest management and that there are periods during each forest rotation (particularly during harvesting and initial stages of regeneration) when not all acres can serve all uses.

PART III

ECONOMICS

10

ECONOMICS AND LAND ALLOCATION AT THE URBAN/FOREST INTERFACE

Perry R. Hagenstein

Two issues with economic content concerning forest land use at the urban/forest interface seem relevant for this symposium. Both concern the increasing impacts of urban development and urban people on forest land. At the heart of these issues is a concern that the market may not adequately reflect all of society's wants and expectations from forest land.

One issue has to do with our ability to produce timber and other resources on forest lands to meet national needs, however these might be defined, or to support regional economic growth and stability. Does urban growth threaten our ability to produce timber, or perhaps other valuable forest resources, either nationally or regionally? The second issue has to do with the value of forest land as open space at the urban/forest interface. To what extent is protection of forest land at this interface needed to ensure its contributions to the value of other land or in providing other values just because of its forested character and because it is undeveloped space?

The assumption behind these issues is that the market for forest land and its uses has some characteristics that limit the extent to which it provides all of the values that society assigns to them. The long time needed to grow timber is one factor that is recognized by the market and that keeps timber production below some biological maximum. Also, the fact that forest landowners cannot capture in the market some of the values produced by their land surely limits production of such values.

These two economic issues are not new to forestry and forest land use policies, although their connection with urban land uses may be. The issue of protection of forest lands to ensure future supplies of timber was cause for national concern long before preserving agricultural lands became the issue that it is today. If some of this sounds familiar, it is because the issues have been discussed for years.

The extent to which either issue is important at the urban/forest interface depends in part on the definition of the interface. An economist's definition may broaden our view of what this is and, at the same time, reduce the degree of our concern. My plan is to discuss the concept of the urban/forest interface first and then examine the two economic issues identified above.

DEFINING THE URBAN/FOREST INTERFACE

Some two decades ago, the head of a forestry department at a major university stated, 'Forestry is not in competition with agriculture, urban and suburban development, or other uses commonly considered to hold higher priority for the use of our land resource' (Spaeth 1960). Robert Healy, in an earlier chapter, discusses forestry as a 'residual' use of land and lists a 'hierarchy' of land uses that has forestry at the bottom.

This common viewpoint would, I believe, be supported by a study of average land values. Average values for forest land for timber production or other uses of the forest are almost certainly less than those for land in agriculture or various residential and commercial uses. But average values are not necessarily the relevant values for comparison.

An economic definition of the urban/forest interface is that point where the marginal value product of the land in an urban use is exactly equal to the marginal value product of the land as forest. As we move away from this margin to land that is more valuable for urban uses, urban values exceed the forest values. But we can also move away from the margin to land that is more valuable as forest. The value of land in forest can exceed the value of that same land in urban uses. One relevant point at which to compare land in forest uses with land in urban uses is at the margin. The values to be compared are marginal values rather than average values.

This definition is difficult to describe precisely on a map. But when we start with the idea of comparing values at the margin, rather than comparing average values, we soon drop the idea of a fixed hierarchy of land values. At the margin, which is where decisions are made, uses of forest land as forest compete evenly with urban-related uses of forest land, whatever they may be.

The value of forest land, whether for timber or for other purposes, varies widely, as does the value of other land. In this sense, the urban/forest interface occurs within a band, possibly a wide band, where uses of forest land with different values compete with urban land uses with different values. Relative land values within the band are determined mainly by accessibility, relationship of uses to other uses, and productivity of the land itself (i.e., site class, slope, aspect, etc.) for various uses. These factors vary for each use. Thus, when they are compared for even a crude breakdown of uses into broad categories such as industrial, commercial, residential, agricultural, and forestry, it is obvious that a large number of points define the urban/forest interface.

None of the definitions seem to capture very well the idea of urban forests and urban forestry. At the recent Second National Urban and Community Forestry Conference, organized by the American Forestry Association, two views of urban forests were most evident. One is that they are composed of trees as individual units within a city or town as usually defined. In a way, this is shade tree management, and arboriculture has more to offer than forestry, as usually defined. The other view is that the urban forest is the entire complex of trees and related biota in an urban area, and should be viewed as a system. Consideration here is extended to the contribution of trees to management of air and water quality and wildlife. Both of these approaches deserve attention. I intend, however, to focus on the economic issues of the interface outside of the immediate urban centers.

Defining what we mean by urban uses is also important for a discussion of relative values in the urban/forest interface. In a region that is heavily urbanized, is there any area not considered part of the urban complex? It is obvious in urbanized regions that urban influences on land values extend well beyond the boundaries of what we usually think of as urban areas. It is also clear that these urban influences are reflected in the prices of forest land.

New England, for example, is both heavily forested and heavily urbanized. Urban areas themselves cover a substantial part of the region, but the total area is still probably less than 10 percent, even in the three southernmost states. But urban influences on land values extend over much of the nonurbanized area. For example, prices of much of the forest land in the region are set in part by the demand for recreational homesites. In this way, the influence of the Boston area extends well into western Maine, central New Hampshire, and western Massachusetts, and that of New York City into western Massachusetts and Vermont.

Within this broad band, urban influences do not extend evenly to all forest land. Accessibility is an important factor, but so too are the character of the land for urban-related uses, its location relative to bodies of water, and its relation to other uses.

Henry Vaux has added another dimension to the idea of the urban/forest interface. He describes it as a political phenomenon (Vaux 1982). He points out that the conflicts in land use at the interface come about because of sharply different views of the social value of forest lands. It seems clear that social values derived from or dependent on mainly urban values are having an impact on forest land uses well beyond the boundaries of urban areas as they are usually defined.

The urban/forest interface thus is a concept with various meanings. The relevant economic interface between uses—the point at which the marginal revenue products of alternative uses are equal—occurs at various places within a broad band of locations surrounding an urban area. The marginal revenue product of each use at each point is determined by such factors as location, access, productivity of the land or alternatives uses, and expected prices. But accepting this definition for discussion here need not preclude accepting other definitions as they may be relevant.

FOREST LAND FOR TIMBER PRODUCTION

The national forests were established because of a national concern, at the turn of the century, about future timber supplies. Forest land was set aside and the timber on it managed conservatively as a hedge against rapid depletion of timber supplies on private lands. This was viewed as a national problem needing a national solution.

A similar argument has been used recently in seeking protection of farmland at the edge of urban areas. It is argued that the unchecked spread of industrial, commercial, and residential land uses threatens both national and regional self-sufficiency in producing food and fiber. Massachusetts, for example, passed a farmland protection law about four years ago in part based on the notion that this would help if truckers' strikes or other calamities were to interfere with food shipments. In reality, the protection program is largely symbolic as far as self-sufficiency is concerned. This concern may be more relevant at the national level.

Can the same argument be applied to forest land? That is, are we in danger of running short of timber, or of other forest resources for that matter, because of the encroachment of urban areas on forest land? Do we face the same problem with forest lands today that many thought we faced some eighty years ago? Does the problem warrant the same kind of approach that is being used in some states that seek self-sufficiency in agricultural crops?

The U.S. Forest Service has long argued that special attention to ensuring future timber supplies is warranted. In its recent assessment of forest and range land in the United States, the Forest Service concluded that meeting 'projected demands for forest and range products requires substantial investments,' presumably beyond those that will be provided by the private sector. At the same time, the Forest Service noted that the nation has a 'huge forest and range land and water base,' but production of most renewable resource products from this base 'is much below what can be attained' (U.S. Forest Service 1980).

Satisfying future markets for timber, according to the Forest Service, does not appear to be severely constrained by limits on the area of forest land. The Forest Service assessment notes that small size, other management objectives, and location limit management for timber in some areas and that some of the forest land is not available for timber harvesting at particular times.

But the Forest Service also notes that, sooner or later, nearly all the timber on lands not reserved from timber harvesting by law becomes available and is harvested.

In a concurrent analysis of the timber situation in the United States, the Forest Service concluded that we can expect relatively rapid increases in the price of softwood timber over the next twenty to fifty years (U.S. Forest Service 1982). The expected price increases indicate an economic shortage of timber. A projected drop in the area of commercial forest land is one of the factors, though not a major one, contributing to the price increases.

The Forest Service projects a 15 million acre decline in the area of commercial forest land from 1977 to 1990 and a continuing drop of about 5 million acres per decade for the next forty years. About half of the decline from 1977 to 1990 is in the national forests, especially in the West, where expected designation of forest land as wilderness will remove it from timber harvesting.

The total area of commercial forest land in the United States apparently peaked in the 1960s, by which time the major shift of land out of agriculture during this century had run its course. As Emery Castle has noted, we can no longer 'look to American agriculture to surrender resources to the rest of the economy to the extent that it has in the past' (Castle 1979). This is as true for land going from farms into forests as it is for labor going from farms into manufacturing. In fact, recent trends in forest land area indicate some shift in the opposite direction. In particular, there has been a considerable shift of forest land into agriculture in the Southeast and the Mississippi Delta. Similar shifts are occurring in other regions as well, although the total area affected does not appear to be as great.

The decline in area of private commercial forest land is projected to be about 7 million acres from 1977 to 1990 and about 6 million acres from 1990 to 2000. This includes the shifts that are expected from forest to agricultural use, which probably account for most of the decline in private forest lands. It also includes shifts from forest to highways and reservoirs, as well as whatever changes take place at the urban/forest interface. The total shift out of private commercial forest land is expected to be something less than 2 percent per decade.

By itself, this expected shift out of forest land is of little consequence to future timber supplies (the amount of timber supplied to the market) over the next several decades. For example, in a recent study of forest policies in the Pacific Northwest, an effort was made to define the effect on expected timber harvests and prices of such a loss of private forest land. First it was assumed that the loss of nonindustrial private forest land would amount to about 5 percent per decade starting in 1980. It was also necessary, then, to make some assumption about the fate of the timber on forest land that is cleared for highways or urban development.

In the Northwest, much of the timber from land clearing finds its way into the market as logs. In effect, this timber substitutes for timber from land that stays in forest. Timber values are sufficiently high that few trees go unused. In the near term, the effect of land clearing on the volume of timber available for industrial uses in the Northwest will not be great. After several decades, however, the impact of such loss of forest land could be significant (Bruner and Hagenstein 1981). In the Northeast, little timber from land clearing has gone to waste since the recent increase in the markets for firewood.

Nevertheless, the Forest Service's conclusion that most timber grown on private lands will be available for harvesting sometime may be unduly optimistic as a basis for setting future timber policies. Two-fifths of the nonindustrial private forest land in the country is in tracts of less than 100 acres each (Jones and Paxton 1977). In addition, substantial acreages of land defined and measured by the Forest Service as commercial forest land are in densely populated suburban areas where timber harvesting for commercial purposes may not be likely. An example is Montgomery County, Maryland, a District of Columbia bedroom suburb, whose 100,000 acres are included in the Forest Service inventory of timberland available for commercial harvesting.

Jones and Paxton present a widespread view on the availability of timber from much forest land held in small tracts. They argue that (1) timber harvesting may be uneconomic on very small tracts and (2) owners of such tracts may have goals that preclude timber harvesting. These are important matters if future timber supplies are a serious national concern. Private forest land in rural areas is being lost for future timber harvesting without a change in its forest cover and even without substantial development. But how important this will be is uncertain, as is the Forest Service assumption that timber on most land will become available for harvesting sometime.

Lloyd Irland, in a recent book on New England forests, speculated on the loss of forest land for timber production. He compared two possible scenarios with estimates of current forest land. His optimistic scenario—optimistic from the standpoint of being able to produce timber—has 21 million acres available for timber production, compared with the current 23 million acres. His pessimistic scenario has only 15 million acres, a substantial loss. Most of the difference is a loss of forest land for timber production to development for recreational subdivisions and small group uses (Irland 1982, pp. 184-85). These are mainly urban-related uses, even if they do not occur within urban areas.

Pressures from urban places and urban people in the broad belts around urban areas have led to a significant fracturing of forest landownership patterns and to an increase in the area of land owned by people whose goals may preclude timber harvesting. Development of land that is now rural in character is continuing. In an unpublished paper, E. H. Kracke estimates that urban expansion over the next twenty years could double the area of developed land, including that in recreational homesites, in southern New England and that this ratio would be even greater in northern New England (Kracke 1978).

This kind of development may have more impact on forests in New England than in other regions because of the large proportion of forest land and the large populations in nearby New York and Boston. But there is little doubt that the same sort of urban impacts on forest land are of increasing importance in other regions as well.

At first glance, then, at least a superficial case can be made that there may be a future shortage of forest land available for commercial timber production. But the question remains: how important are these impacts on the forest land base in terms of national or regional timber supplies? Are forest lands plentiful enough that removing a portion of them from the timber supply base will not have a major impact on future prices or harvest levels? Two recent studies help to answer these questions.

The Forest Industries Council estimated the increases in timber supplies that would occur if private lands were managed more intensively. The estimates were made using the Forest Service Timber Assessment Market Model (TAMM), which was developed to help prepare the periodic programs required by the Renewable Resources Planning Act (RPA) (Adams and Haynes 1980). The FIC estimated that net growth of timber in the United States could be increased about 50 percent over current levels if investments that would return at least 10 percent in real terms were made in timber production (Forest Industries Council 1980). Even greater increases in timber growth could be had if 4 percent, instead of 10 percent, were the acceptable cutoff rate of return.

These results suggest that the area of forest land does not have to be the factor that limits future timber supplies. That is, it is apparently economically feasible to offset losses of forest land to timber production with more intensive management if the goal is to provide some specified level of timber production.

The other study was part of the economic analysis of forest policies in the Pacific Northwest mentioned earlier (Bruner and Hagenstein 1981). Clark Binkley examined the economics of investing in intensive timber management for alternative future timber price forecasts. He found that current state forest practice laws in the Northwest make intensive management economically advisable even for the policy scenarios that result in minimal increases in future timber prices (Binkley 1981).

These forest practice laws were passed to avoid more stringent water-related laws being imposed. They require, in effect, that timber harvesting must be followed by regenerating the site to suitable timber species, which often means planting. Binkley's finding was that once the investment has been made in planting, additional investments in precommercial thinning and fertilizing are economically efficient on practically all sites and for timber price forecasts that range from rapid to slow price increases. Thus the stringent requirements in the forest practice laws that require replanting following logging, and which were adopted for reasons unrelated to concerns with timber supplies, have a considerable impact on the economics of timber production.

In sum, using assurance of future national timber supplies as a rationale for resisting urban encroachment on forest lands does not appear to be warranted. Even if one accepts the proposition that future demands for timber will somehow outrun available timber supplies, intensification of timber management on the remaining forest lands, which would still be extensive, can easily offset any likely losses of forest land.

To continue the parallel with the farmland preservation issue, can a plausible argument be made for protecting forest land for timber production from a regional standpoint? Like most agricultural products, forest products move readily in national markets. However, while interruptions in the supply of most forest products would be inconvenient, they would not cause the same concern as interruptions in the supply of food. In this sense, at least, there is little reason to protect forest lands to ensure regional availability of forest products.

Protecting forest land as a regionally significant base for employment and income may be another matter. If this is important anywhere in the country today, it is in the Pacific Northwest. During the 1970s, average annual direct employment in the lumber, plywood, and pulp and paper industries accounted for 5.5 percent, 8.7 percent, and 4.9 percent of the employment in, respectively, Idaho, Oregon, and Washington. By the 1990s, and with a continuation of current federal and state forest policies, these percentages are expected to fall to 2.9, 4.2, and 2.5 percent (i.e., a decline of about 50 percent over two decades) (Bruner and Hagenstein 1981, p. A-19).

In this same study of forest policies in the Pacific Northwest, we simulated a loss of private forest land. The effects of the loss on timber harvests and prices were estimated. The simulated loss in the 1980s in western Oregon and western Washington was 20 percent of the nonindustrial and 5 percent of the industrial forest land. The results of this simulation were about a 7 percent decrease in timber harvests in the region in the 1990s relative to a base simulation and a 6 percent increase in timber prices (pp. 6-76 to 6-83). The decrease in harvests translated into a parallel decrease in timber-based employment.

Whether such a loss of forest land to timber production should be perceived as an economic loss to the region depends on the alternative uses to which the forest land would be put and the returns to forest landowners. It is important to note that the decrease in timber harvests that result from the simulated loss of forest land is accompanied by an increase in timber prices.

In the case described above, timber owners as a group are slightly less well off from a loss of forest land than they would be without such a loss. If timber prices go up more than total timber harvests go down, the result is a transfer of income from purchasers of timber products, mainly outside of the region, to landowners in the region. Even then, the increased income of timber owners would not necessarily benefit the region. Capital is a mobile factor of production. It will move to the most attractive investment opportunities, even if they are outside the region.

Keeping forest land in timber production will benefit workers and processors in the region. Its likely impact on the income of timber owners and on the region's wealth, however, is not at all clear. The case for protecting forest lands to support a regional economy is tenuous.

FOREST LAND AS OPEN SPACE

Like farmland at the edge of urban areas, forest land has value as open, undeveloped space. At the root of efforts to protect farmland is a concern not just for crop production but also for protecting open space and preserving flexibility in resource use (Castle 1980). Similar reasons can be offered for retaining forest land at the urban/forest interface. The problem is in deciding on the extent to which intervention in the market for forest land at the interface is warranted to retain open space and preserve options for future land uses.

There is some evidence that the open space values of forest land are already considered in the market for forest land. For one thing, there are still substantial areas of forest at the urban/forest interface, at least in forested regions, without direct intervention in the market for land.

One example of this is in the large size of residential lots, especially in the suburbs, in most urban areas. For whatever reasons, perhaps because of the American frontier heritage, homeowners like to have more land than is needed just as a building site. The cost of providing this open space for the landowner is part of the total cost of the physical environment that is valued by the owner.

Most of this kind of open space is bought at a price that far exceeds what would be paid for the land if it were to be used for producing timber or wildlife, especially if the land is close to an urban center. Thus the market is recognizing some value other than the value of the land for forestry and, indeed, other than its value for building sites. The market provides at least part of the forested or tree-covered open space that people believe desirable.

As the distance from urban centers increases, the idea of forested open space having a value that is recognized in the market seems less likely. But land prices suggest otherwise. Last spring, I took a brief look at forest land prices in southern New Hampshire in an area some 50 to 90 miles from Boston. Prices reported to me were from $450 to $500 per acre for ordinary forest land in tracts of 100 to 500 acres. The descriptions usually included something on the attractiveness of the site, indicating that the prospective sellers have some idea of the value of the land as open space, probably in relation to large-lot homesites. Again, the market price of forest land in this extended definition of the urban/forest interface appears to exceed by two or three times its value for timber production.

Irland and Field (1980) have reported on the impacts of suburbanization on timber growing in Connecticut. They have found that the three counties in Connecticut with the most rapid growth in this century are the most urban in character and also have increased in area of forest land. The significant point is that land prices in these counties are high, but lot sizes outside the urban centers are large. Owners have paid for something that they value, and it is unlikely that they are paying for the land's timber production potential.

Recognition of open space values of forest land in the market is often buttressed by government action that ensures that off-site, or external, impacts will be considered. Restrictive zoning in the form of large minimum lot sizes guarantees that property values that depend on the sense of open space will not be lessened by the actions of a few property owners. Widespread use of such zoning, especially in areas where land prices are high and open space is assigned significant value in the market, is evidence of its acceptability.

Restrictive zoning presumably results in the public acquiring something of value to which it previously did not have a right (Sax 1971). But in the case of restrictive zoning that leads to protection of open space, it is not clear that the public acquisition of something of value has been at the expense of the landowners whose use of land is being restricted. That is, the owners themselves accept the restrictions and thereby increase the value of their properties. Zoning is responsive to the realities of the market for land as well as to some planner's vision of the ideal arrangement of land uses.

Some useful research could be done to measure the extent to which the value of forest land is recognized by the market at the urban/forest interface. If forest cover adds value to property, it should be apparent by comparing the records of town or county assessors with the physical condition of individual properties. Similar comparisons of property values between towns with different degrees of forest cover should show the same results. Also, the extent to which homeowners invest in improvements to their property could be compared with the forest cover of their lots. I am convinced, although I have no data, that forest cover adds to value.

Fee acquisition of land by government is also used to protect open space in a way that may be responsive to market values of land. My town of Wayland, some twenty miles west of Boston, provides an example. The population density of Wayland is about 800 people per square mile. The town has had an active conservation land acquisition program for the past two decades. About 7 percent of the town's area has been acquired by the town for conservation land and has been removed from the tax rolls. Despite the obvious costs to the town's voters of such acquisitions, they have approved appropriations for additional conservation land at town meetings year after year. Neighboring towns have similar programs.

Land values in Wayland are among the highest in the Commonwealth for towns with its population density. Some of this high value is attributable to the relatively high proportion of open space in the town. There is little doubt that the voters believe they are voting for higher property values when they vote for more conservation land. The tax money is not being spent for wholly altruistic purposes.

One of the arguments used at town meetings to support buying more conservation land is that it will, in the end, lower tax rates. The idea is that taking buildable land off the tax rolls will limit the possible increase in school-age population. This, it is argued, will lower building needs and budgets for the school system. There can be little doubt that the voters, in accepting such arguments, are voting with their pocketbooks.

Especially in recent years, when the school-age population has been falling, voters are probably looking at more immediate effects than long-run impacts on the school budget. The first economic effect of conservation land acquisition is an increase in the tax rate because of (1) the cost of land acquisition and (2) the loss of land on the tax rolls. These are the direct costs that have been accepted by the voters. These are costs that I believe they see as being offset in good part by increased property values that are shared by landowners in the town.

In sum, forest land has value as open space and as a backdrop to other uses of property at the urban/forest interface. A significant part of this value is recognized in the market for land, even where collective action is needed to ensure that the values can be captured. Two examples of this kind of collective action are restrictive zoning and acquisition of open space by local government. Economic values of such action are reflected directly in increased values of private property.

CONCLUSIONS

I have presented two arguments. First, protection of forest land at the urban/forest interface is not needed to ensure future timber supplies. This approach makes little sense from an economic perspective. Second, much of the economic value of forest land as open space at the urban/forest interface is reflected in market values and in actions of local government that have an economic basis.

Is there an economic issue, then, concerning forest land at the urban/forest interface that warrants public policy attention? The answer is yes. Some off-site, or external, values of forest land as open space are not reflected in the market.

That such values exist seems obvious from the attention given to them in literature, the arts, and the media. It seems equally obvious that these are values held to be important by a broad cross section of people that extends beyond the suburban and rural residents who can provide through ownership at least part of the forested open space they desire.

Although marginal market values of forest land as forest and in some urban use will be equal at various points in the band around suburban centers, average values of the land in forest increase relative to average values in urban use as the distance from an urban center increases. Thus, if there are off-site open space values in forest land that the public wants to capture, the cost of doing so will be highest close to urban areas. This is unremarkable. But the implication for public action is important.

Efforts to protect off-site values by modifying the market for land are most likely to be successful well away from urban centers. For example, adjustments in property taxes, such as assessing property for its current use rather than for its highest and best use, are more likely to be effective in keeping land in forest where the average values of land in forest or in developed uses are reasonably equal. This is most likely to be the case in rural areas.

Direct action, such as public acquisition of full or partial title, is more likely to be needed to protect open space off-site values of forest land in areas close to urban centers. Effective overall programs must be applied differently at different points of the urban/forest interface. Costs of such programs per acre of forest land will decrease as the distance from urban centers increases, but it is likely that per acre benefits will also decrease with distance.

In sum, the market for forest land at the urban/forest interface provides much of what society wants and expects from the forest. I have claimed much for the role of the market in allocating forest land efficiently. I have claimed little for the role of the market in securing a fair distribution of the benefits and costs of these allocations, either today or among generations. Intervention in the market for forest land at the urban/forest interface may be warranted where questions of distribution are paramount.

REFERENCES

Adams, D. M., and R. W. Haynes. 1980. The 1980 softwood timber assessment market model: Structure, projections and policy simulations. Forest Science Monograph 22.

Binkley, C. S. 1981. Optimal timber management intensity in the Pacific Northwest. In: W. E. Bruner and P. R. Hagenstein, eds., Alternative forest policies for the Pacific Northwest.

Bruner, W. E., and P. R. Hagenstein, eds. 1981. Alternative forest policies for the Pacific Northwest. Prepared for Forest Policy Project, Washington State University, Pullman. May be obtained from NTIS PB 82 109125.

Castle, E. N. 1979. Food, energy, and land: Abundance or scarcity? Paper presented at Annual Meeting, American Forestry Association, Charleston, S.C., October 29.

———. 1980. Every body has to be some place: The role of land in a changing economy. Paper presented at Pacific Northwest Regional Economic Conference, Portland, Oregon, May 3.
Forest Industries Council. 1980. Forest productivity report. National Forest Products Association, Washington, D.C.
Irland, L. C. 1982. Wildlands and woodlots. University Press of New England, Hanover, N.H., and London.
Irland, L. C., and D. B. Field. 1980. Impacts of suburbanization on timber growing: A case study of the State of Connecticut. Connecticut Woodlands 45(2):16-24.
Jones, R. E., and J. S. Paxton. 1977. The 296 million acre myth. American Forests 83(11):6, 8.
Kracke, E. H. 1978. State growth policies in New England: Their status, successes, and shortcomings. New England Natural Resources Center, Boston, June. Draft.
Sax, J. L. 1971. Takings, private property and public rights. Yale Law Journal 81:149-186. Reprinted in: B. A. Ackerman, ed., Economic foundations of property law. Little, Brown, Boston, 1975.
Spaeth, J. N. 1960. In: Land Economics Institute, Modern land policy. University of Illinois Press, Urbana.
U.S. Forest Service. 1982. An analysis of the timber situation in the United States: 1952-2030. FRR-23. U.S. Government Printing Office, Washington, D.C.
———. 1980. An assessment of the forest and range land situation in the United States. FS-345. U.S. Government Printing Office, Washington, D.C.
Vaux, H. J. 1982. Forestry's hotseat: The urban/forest interface. American Forests 88(5):37, 44-46.

11

MARKET AND NONMARKET INFLUENCES IN URBAN/FOREST INTERFACE CONFLICTS

Roger A. Sedjo

Perry Hagenstein has provided us with a lucid, concise discussion of the problem of the urban/forest interface. He has presented a useful view of the interface margin in both physical and economic terms. Why be concerned about forest lands at the margins of the urban areas? Hagenstein gives two reasons: (1) concern about long-run regional and national timber supplies and (2) concern about the amenity values that 'open spaces,' in this case forests, provide to the citizens of a location.

Addressing first the issue of long-term timber supply, Hagenstein points to the role that technology and intensive forestry are likely to play in the production of future timber supplies. We have seen much of this before in agriculture. While the capacity of the land to meet the demand for food has been a recurring public concern since before the time of Malthus, actual experience has been quite positive. New technologies that raise yields are substitutes for land. In agriculture the effect of these technologies on yields has been phenomenal and has resulted in rapid expansions in food supplies from a limited land base (Heady 1982). I believe that management and technology can also be readily substituted for land in forestry. Although clearly still in the primitive state, silvicultural practices such as planting, thinning, and fertilizing have demonstrated positive effects on yields (Hyde 1980). Furthermore, genetic improvement in forestry is still in its infancy and shows tremendous potential for increasing yields. The implication is that in the future greater portions of our supply will be produced through intensive forestry using improved technologies on what will be relatively modest land bases. Therefore, with Hagenstein, I do not believe serious concern is warranted over the long-term effect on timber supply of the modest reductions in the forest land base that are occurring.

There is no reason to believe that the unregulated private market could not adequately provide socially desirable levels of timber production. Timber is a commodity, and like other commodities there are market incentives associated with its production. We have a large and diverse forest industry with a demonstrated ability both to harvest and to grow trees. Its future as an industry depends critically on its ability to anticipate and respond to societal needs for timber. Should the long-run timber supply be perceived in jeopardy, the returns to forest production would rise and the industry would see its self-interest dependent on finding ways to expand future production. One method would be to shift land at the margin from other uses to forestry. Nonindustrial private forests could also be expected to contribute. While it is true that many nonindustrial private owners have nontimber goals, it is also true that others include timber in their land use objectives. Economic incentives would influence the extent to which the responsive ownerships would move toward commercial production.

On the issue of forests as 'open space,' Hagenstein notes that there is a problem 'in deciding the extent to which intervention in the market for forest land at the interface is warranted.' He gives several examples of how the market mechanism allows for 'open space' values to be captured by individual owners. Larger lots for private residences which allow for a forestlike environment are common in many places, including, as he suggests, the suburbs in major metropolitan areas. He notes that zoning can accomplish this end and may generate socially desirable results. In effect, he argues that zoning may be a relatively low cost way of internalizing certain externalities. Although this may be true for some zoning ordinances, one must be careful with this argument for two reasons. First, as Hagenstein notes, some external values may not be reflected in the market or in the actions of local government. Second, for zoning to internalize externalities, landowners must have relatively homogenous situations and preferences.

My major criticism of Hagenstein's paper would be that insufficient attention is devoted to the inherent elements of conflict associated with the urban/forest interface. I tend to accept Henry Vaux's observation that the interface is really about the conflicts of values that exist between different groups each with a somewhat different set of perceptions of how society ought to use its forests and forest lands (Vaux 1982). With this view, the issue includes not only questions about land uses at the periphery of the urban areas but also questions as to alternative uses of forest lands regardless of location and questions as to the appropriateness of various restrictions on private forest lands, particularly those that provide amenities for the public.

One institution available for resolving these conflicts is the free market. In the extreme paradigm of that institutional context, land could be used for whatever purposes the owner deemed appropriate. Conflicts at the urban/forest margin would be resolved by the market. The individual who purchased the land could use that land for purposes of his own choosing. Development would replace forests when developers were prepared to bid land away from owners with forestry objectives. Conversely, if the forest values to the owner, whether commercial or amenity, exceeded the development values, the land would be expected to remain in forests.

The economic argument against unconstrained market solutions relates, of course, to the existence of externalities. A straightforward example of an externality would be downstream pollution and silting that might result from a logging operation. In such a case it is often argued that government intervention is justified.

The question arises whether this same rationale should be applied to forest amenities. For example, should timber harvests be restricted on private lands along roadsides to preserve visual amenities to the traveler? If harvest is prohibited through regulation, the timber owner loses the values associated with the timber harvest. Should the owner then be compensated for this loss, and if so, by whom? Recent court decisions have not generally provided compensation for financial losses due to regulation (Johnson 1983, Bosselman 1983).

Analogous conflicts exist in the management of public forest lands. For public lands, of course, the market is not expected to dictate land usage. However, Krutilla and Haigh have argued (1978) that the legislative mandate is that the forest be utilized for its highest value where both market and nonmarket outputs are given their appropriate social values. But the absence of well-developed markets for many of the outputs often makes it difficult to determine the level and mix of outputs from public lands that would maximize the social output of the forest. Furthermore, the decision ceases to be determined by economic considerations and instead is determined by some variation of the political and bureaucratic processes. Interest groups find it effective to attempt to influence this process to achieve their particular goals. Depending on one's point of view, the outcome may be seen as producing either socially desirable results, which incorporate nonmarket values, or results that are consistent with the preferences of a particular interest group but not generally socially optimal.

While the rationale for government intervention in the operation of markets and, indeed, the fundamental rationale for public ownership of vast quantities of forest lands are founded in the concept of market failure, disillusionment with the results of governmental intervention has resulted in a growing counterview that 'there is certainly no *a priori* reason to believe that market failure is any more pervasive than governmental failure' (Anderson 1982). This school of thought argues that market results, even when imperfect, will generally be preferable to intervention either through regulation or governmental ownership.

In this regard, let me digress just a bit to discuss a provocative new thesis advanced by Mancur Olson in his recent book, *Rise and Decline of Nations* (1982). Olson's thesis, simply put, is that the longer nations are politically stable, the more interest groups they develop and the worse they work economically. Olson uses this thesis to explain, for example, the rapid economic growth experienced in the post-World War II period by Japan and Korea compared with the low growth of the U.S. and Britain. A manifestation of this common-sense phenomenon in the forest resource arena is found in the urban-rural conflict in forest values that Vaux observes and the method by which this society deals with that conflict. Through time, the market has been increasingly disqualified from its role as a resource allocator on grounds that it fails to account properly for externalities and other societal values. Its place has been increasingly taken by politically instituted systems for governmental intervention that typically involve interest groups vying with one another to achieve their individual specialized aims. The more that interest groups achieve their specialized objectives, the more they displace the market as a mechanism for making these economic decisions. Casual observation suggests the growing role of interest groups in recent years in resource decisions in forestry and other natural resource areas. If Olson's thesis is correct, the net effect is poor decisions from the broad social point of view and a reduction in the economy's overall efficiency. A maddening aspect of Olson's thesis is that it is the very political stability of a country that allows interest groups gradually to achieve power that then gradually weakens the economy.

To conclude, I agree with Hagenstein that the concerns expressed about forest land availability to meet long-run demands for wood are often overstated and are certainly no cause to attempt to modify the allocation between forest and other land uses. I also agree that the market is in fact incorporating many of the amenity values of forest land into the price of that land. I am, however, much more concerned than he appears to be by the substantial conflicts at the urban/forest interface. I am particularly concerned at the growing dismissal of the market as a mechanism for economic decision making due to its imperfections, alleged and real, and the substitution of a system of interest group dynamics and bureaucratic management.

REFERENCES

Anderson, T. 1982. Institutional underpinnings of the water crisis. Paper presented to the Land Use and Resource Development Conference, Big Sky, Montana, August 25-29.

Bosselman, F. P. 1983. Comment. In: R. A. Sedjo, ed., Pressures on U.S. forest lands. Forthcoming (Resources for the Future).

Heady, E. 1982. The adequacy of agricultural land: A demand-supply perspective. Chapter 2 in: Pierre Crosson, ed., The cropland crisis: Myth or reality? Published for Resources for the Future, Johns Hopkins University Press, Baltimore.

Hyde, W. F. 1980. Timber supply, land allocation and economic efficiency. Published for Resources for the Future, Johns Hopkins University Press, Baltimore.

Johnson, M. B. 1983. Regulation of private forest lands: The taking issue. In: Sedjo, ed., Pressures on U.S. forest lands. Forthcoming (Resources for the Future).

Krutilla, J., and J. Haigh. 1978. An integrated approach to national forest management. Environmental Law 8(2):373-415.

Olson, M. 1982. The rise and decline of nations. Yale University Press, New Haven.

Vaux, H. J. 1982. Forestry's hotseat: The urban/forest interface. American Forests 88(5):37, 44-46.

12

INDIRECT IMPACTS AND INEQUITIES IN URBAN/FOREST INTERFACE ECONOMICS

Clark Row

I greatly enjoyed Perry Hagenstein's excellent paper on the economics of land use in the urban/forest interface. Like Roger Sedjo, I agree that there is little likelihood that development of forest land to urban uses in interface areas will by itself substantially decrease the aggregate supply of forest products to the U.S. economy. Nor will increases in the prices of forest products caused by changes in demand and supply retard the expansion of the interface around most metropolitan areas. Because desirable sites are an important component of housing attractiveness, developers willingly pay prices for land that greatly exceed the value for growing timber on all but the most productive forest sites.

Direct use of forest land for residential and urban uses, however, is only part of the impact of urban land development. Indirect impacts may be far greater: the increased subdivision of all land to parcel sizes below thresholds for efficient timber production, speculative land acquisition for future development, and pressures for zoning and other restrictions that prohibit or effectively discourage management investments, and even timber harvesting, on land not currently developed. Each of these indirect impacts justifies additional comment.

The decline in land use (and value) for timber produced on small tracts or parcels has two causes. The first is the high fixed management cost per tract for planning, treating, and moving equipment to the site. As demonstrated by several studies I have made, the more equipment-intensive the treatments, the higher the fixed costs. The second cause is the decrease in timber prices received for timber cut from small tracts. Effects of tract size have been found in numerous stumpage price studies, especially in the South. The relationships found there also apply in the Pacific Northwest and other regions. The loss in stumpage value is apparently caused largely by high fixed costs of moving loggers and their equipment to the site. The net effect is a sharp reduction of net worth of land for timber production as parcel sizes decrease; and at threshold sizes—varying from a few acres to 50 or more—it becomes negative. In such cases timber growing is not economical.

Owners holding land for speculative purposes seldom are interested in managing the land to produce forest products, according to many owner surveys. The reasons are simple. Owners expect to sell their land before they recover investments in timber cultural treatments and often do not harvest timber because it might adversely affect the attractiveness of the site. A remarkably low proportion of persons receiving cost-sharing payments under the Forest Incentives Practices program convert their land to nonforest uses within the first five years of treatment. Speculative owners do not even apply for the payments.

As others have mentioned, a mixture of residential and forest lands creates pressures to limit activities that cause unsightly or otherwise distracting changes. Use of pesticides often arouses health fears. Logging, with its noise and heavy truck traffic, may also be restricted. In New Jersey and other heavily settled eastern states, logging has been banned by many municipal or township ordinances. The same environmental impacts are caused by almost all types of construction, but these appear temporary and do not recur on the same tract.

Reduced timber harvests within the urban/forest interface may cause another effect that has not been mentioned. If sawmills cannot obtain sufficient timber within a reasonable transportation radius, they close down or sell their valuable sites to other industrial concerns. The loss of business mills reduces competition, further reducing timber prices and the incentives to manage land for timber production. Loss of local processing facilities may also sharply reduce processing of timber removed from tracts under development.

The net effect of these indirect impacts is unknown. But close study of both direct and indirect effects is needed before it is concluded that the expanding urban/forest interface has no significant effects on timber supply.

Analysts associated with forestry, such as Perry, Roger, and myself, as well as most contributors to this symposium, tend to emphasize the impacts of the expanding forest interface on the markets and production of timber and other forest products. The consensus would allow most land development for urban uses to occur under market principles within broad land use zoning and regulation. In the land market, low land values of most forest and agricultural land relative to values for residential and urban uses make possible low density land development within the urban/forest interface.

An equally important public policy question concerns the impacts of land development on community services. Attention should be given to the rationale for planning and regulation to reduce the costs of public services, hence total future costs for all local area residents. This concern is most appropriate in an age in which urban dwellers are expecting higher levels of education, fire and police protection, recreation, and transportation services and the extent and cost of local government activities are causing significant political repercussions.

Community services further illustrate the powerful effect of economies of size and scale. The most important determinants of the per unit cost of community and utility services are the extent of use, the number of users, and the investment in distribution lines. Length of distribution lines affects both the initial investment in facilities and the annual maintenance and operating costs.

One only needs to mention the wide variety of public and utility services to appreciate the effects of decreased residential and industrial concentration. Important is the increased reliance on automobile transportation for all in-town movement, including trips to shopping centers from homes that were once in walking distance. Road, water, sewer, power, telephone, and gas networks are expanded. Mail service becomes more costly. Response to calls for fire and police protection is less timely, and more expensive to provide. Another example with large social consequences is that forty years ago only truly rural children were located so far from schools that they had to take buses. But now in most states a majority of students take school-provided or public transportation.

The free market mechanisms do not fully or equitably distribute these costs. They are also not completely reflected in the tax or utility rates, and thus are not capitalized into the land values. Part of the costs are indeed borne by the developers and passed on to the purchasers of the residential and industrial property. These include the direct costs of street, water, sewer, and utility lines within the developments. Since these costs are imbedded in the property values, they raise the cost of housing and industrial plants to some extent. Although home buyers obviously believe the advantages of dispersed locations are worth the costs, social and economic segregation is increased.

But a large share of the cost of services to locations within the urban/forest interface is borne by the public at large—residents of older districts and neighborhoods as well as the new. These are the capital and operating costs of major roads and trunk distribution lines, and the increased operating costs of other services. When the costs are paid from tax revenues, as police and fire protection costs are, property and other tax rates are increased. When the costs are covered by user charges or utility rates, as is increasingly the case, distinctions in charges are rarely made according to neighborhood or density of development.

Thus residents of older, more compact neighborhoods bear much of the costs of the urban-interface expansion. The failure of the owners of property within the urban/forest interface to pay fully for the increased costs they impose tilts the economics of land use toward lower prices of new houses, higher land values, and consequent faster enlargement of the interface than would otherwise prevail.

These diseconomies of community services are thus a concern for local planners, zoning officials, and land use agencies. They and the related inequities are strong arguments for land use regulation and differential changes for public services.

PART IV

PHYSICAL AND BIOLOGICAL DIMENSIONS

13

SILVICULTURE AT THE URBAN/FOREST INTERFACE

David M. Smith

As the general nature of this symposium makes clear, urban forestry differs from other forms of forestry mainly in the social dimension. The fact that it is people oriented is not the difference, however. All forestry is concerned with meeting the needs of people first and trees second. The difference is that in urban forestry there are more people with a greater variety of demands, more harmful effects on the ecosystem, and less understanding than is the case in the rural areas to which forestry is often relegated.

Some years ago someone asked the late Harry M. McKusick, state forester of Connecticut, whether an urban forester really needed to know much about silviculture and that sort of thing because it was all so much a matter of people problems. McKusick's swift and sober response was that the urban forester had to be 'the best silviculturist of all' because the challenges were more numerous, diverse, and difficult than they were out in the countryside.

The urban forester seldom knows which problem is going to hit next and works in an environment that is full of them. If, on the other hand, one is charged with growing pulpwood and sawlogs in an industrial forest, the objectives have a way of being few and self-defining. This means that urban forestry ought to be one of the best fields for foresters who are, in the long run, esteemed most for the problems they solve and not for how well the trees grow. Physicians do not necessarily thrive in healthy communities, and it is worth noting that the urban forest is full of patients needing cures. The urban forest has opportunities as well as maddening problems. Foresters should not be put off by those who know only what cannot be done.

The limits of urban forestry are somewhat vague but probably range from the management of roadsides and street vegetation at the smallest scale to water-supply lands at the largest. One important component is or ought to be the suburban residential forest. The territory is broad enough that I can only present some ideas and observations that come mostly from activity in the city-studded forest that is southern New England.

If there is anything physical or biological that urban forests have in common it is the high incidence of damaging effects from the impact of large numbers of people. Anyone in urban forestry must be knowledgeable about the ways of dealing with all manner of damage to soil, vegetation, and air by biotic and nonbiotic agencies. The indignities that are routinely and unthinkingly inflicted on soils in an urban environment are often of the sort that would cause many foresters to close down a logging operation in the woods. One of the most helpful kinds of information that foresters can bring to the city from the wildwood is knowledge of soil conservation.

Soil damage, toxic chemicals, and other urban phenomena expose trees and other vegetation to biotic pests that may seldom be seen in the more rural forests. Furthermore, the urban environment is often harsh enough to make the vegetation less able to recover from the effects of damaging agencies. In other words, in the urban environment susceptibility to attack is high and so is vulnerability to damage.

Dealing with these matters is not so much the application of certain general principles as it is the ability to analyze problems and apply creative imagination to their solution. It may help to describe a few cases. One involves a forester named Hans Bauer who spent his whole career managing the vegetation along the state highways of Connecticut. He started in 1924 and retired about twenty years ago. Being a good botanist, he proceeded on ecological principles. One of these was that if one grows plants not in soil but in the parent material of highway cuts and fills, one goes to see what natural plants colonize the banks of borrow pits. If a native plant would serve, that would be fine; if an exotic from the similar climate of Japan or China worked better, it would be all right to use the exotic.

Bauer also perceived that it was hard for woody plants to get started in parent material that had no organic matter. Consequently he devised equipment to distribute hay and even woodpulp mixed with grass seeds over the surfaces to be revegetated. He also had an aesthetic objective. At the time he started, Connecticut still had many naturally reforesting abandoned fields, so the shrubs and other vegetation that he planted on the road margins blended with the adjacent landscape. Even today as I drive along Connecticut highways I cannot tell which junipers or sweet ferns were planted by the disciples of Forester Bauer and which were brought by birds or wind. There are enough other ways of handling such problems that I can tell when I have crossed into Massachusetts.

I also work surrounded by the effects of the policies of the late George Cromie, who was city forester of New Haven, Connecticut, about sixty years ago. New Haven still calls itself the Elm City, having overlooked the fact that Dutch elm disease did away with almost all of the elms before 1960. Cromie had anticipated, even before 1920, that something horrible would ultimately happen to the elms, if only because they formed a single age class getting older. He therefore planted various maples under the elms, and there are many miles of streets that would be treeless today had he not done so. He did not know about Dutch elm disease when he initiated the policy.

Unfortunately he did not anticipate everything. Maples are highly vulnerable to salt damage, and in many instances salt is used to combat icy streets. Furthermore, too many layers of asphalt were added to streets without raising the curbing. This caused much salty water to spill over the curbs around the tree roots. The result is attrition of maples. The modern manifestation of the Cromie policy is the planting of a diversity of tree species. One is pin oak, which is adapted in nature to hard-bottomed clay swamps and also to the bad soil conditions along city streets.

URBAN WILLINGNESS TO SPEND MONEY

It would be easy to leave the impression that urban forestry is the repair of one disaster after another. Actually there are some aspects that make urban forestry easier than rural forestry. Usually there are no problems of access to the forest; ordinarily it is too accessible. The urban society is usually affluent and does not expect its local forests to show a profit from commodity production. It thinks of spending dollars where rural forestry interests would spend pennies. While I do not advocate shoveling money out the window, it does help for foresters to remember that the urban clientele is not necessarily addicted to penny-pinching solutions and may even hold them in contempt. While this is certainly not always the case, it is well to anticipate that the financial criteria will not be the same as those of an industrial timber-production enterprise.

There is at least one activity in which urbanites use their power to spend money on trees unwisely. That is in the prophylactic spraying of shade trees with insecticides regardless of whether any pest attack is imminent. The same people who complain bitterly about use of pesticides far off in the woods will blithely hire some 'tree protection' firm to spray trees that are in close proximity to many innocent bystanders. I am not saying that the pesticides themselves are very dangerous but only that their application is often needless. Furthermore, many slick operators are taking large amounts of money under false pretenses. It is gratifying that few of them are professional foresters.

The best tactics of pest management to apply in urban forestry are, as elsewhere, simply those of avoidance and evasion. We can stay out of much trouble merely by matching species of plants to the environment in which we place them. It is only inviting trouble to plant some exotic on a city street merely because it is being pushed by the nursery trade or grows fast if planted on a riverbank ten blocks or 700 miles away.

Urban interests are also prone to spend money unwisely on instant results. Since it is not really understood that trees grow, there is a tendency to want to plant large ones in place. This has the incidental effect of confirming that trees don't grow, because these trees often stand around convalescing for years without growing, even if they survive. Mainly it costs a lot to move the trees and keep them alive. Worse yet, they are often planted in spots where, if they finally grow, they will encroach on utility wires, or plug sewers, or drop leaves in the wrong places. Sometimes the growing tree can cause problems that call forth observations of the sort made by a famous city boss of the old style. When beset by constituents who opposed cutting trees for a sidewalk, he made the classic pronouncement, 'Da place for trees is in da woods!' The best justification we have for planting large trees in urban areas is that it keeps foliar buds and twigs high enough that they don't slap people in the eyes and are beyond the reach of urchins. Otherwise small trees do better and are much cheaper to establish. It is also necessary to anticipate that they will grow, and not always as we would want.

AFFINITY OF CITIES AND FORESTS

Urban forestry does not have well-defined geographical limits. It is difficult to state what is urban, suburban, or even rural. The definitions get all tangled up. This is partly because good rainfall sometimes nourishes both forests and cities. In terms of population distribution— that is, the proportion of the population living in urban areas—the most urbanized states are apparently those of the Pacific Coast. In this sense, Australia is the most urbanized nation in the world. It is also paradoxical that the most densely populated U.S. states are New Jersey, Rhode Island, Massachusetts, and Connecticut, yet there are few other states that are more nearly covered up with trees. In fact, if one examines such places as the Pacific Northwest, Australia, and the Northeastern Megalopolis in detail, it is seen that some forests grow where there are many people.

Sometimes the attitude of foresters about urban influences has been that if we can run far enough out into the wildwood nobody will catch up with us and we can 'practice forestry' the way it should be done. It is perhaps beside the point that we can't hide even there. If we get too far, we find some national forest roadless areas, with devotees of wilderness preservation lurking in the underbrush ready to sue us about an environmental impact statement. If it is a nice, quiet industrial forest, there is a risk of being sold out by the gnomes of Wall Street. Since there is no place to hide from urban influences, it is best to go out and meet the urbanites boldly with the most resourceful plans that we can muster.

We can, as usual, lift some lessons from overseas. The Germans have generally been there before us. In fact, almost all of the forestry that goes on today in heavily populated West Germany seems, even in the estimation of the foresters who conduct it, like urban forestry or 'managing parks.'

Two important lessons can be learned from the Germans' very real success. First, the costs of intricate management and careful logging are so high that it is necessary to focus on growing the most valuable timber possible. This is the only way one is going to pay the costs and still make a profit. It does help to have a timber-deficit economy conducive to high prices. Such products as pulpwood and fuelwood are grown but only as a means to the end, being worth little more than they would be in this country. If growing such low-priced material pays in any country, it is only far out in the rural areas.

The revival of the urban American market for fuelwood can play a useful role by providing the thinnings and improvement cuttings necessary to grow valuable trees in urban forests. Urban areas are woefully deficient in local fuel supplies and cursed by the pollution hazard that looms large for almost any kind of fuel source. While the fuel supply from urban forests would be only a drop in the bucket, almost every bit of it is accessible, nearby, useful, and, if burned properly, almost free of pollution hazard. The economies of scale that dictate that pulp mills be large do not apply to most ways of using wood as fuel. The best possibilities lie in small factories on the suburban fringe where there may be enough wood within 15 miles. It would not be well to rely completely on weekend woodcutters and home heating at one extreme or on the 25-megawatt electrical generating station at the other. The harvesting and burning of wood as domestic fuel often suffers from various inefficiencies, although it may be less costly than golf as recreation. The electrical generating station has roughly the same kind of supply problems as chemical pulp mills; experience shows that they can be solved most easily rather far away from cities. In any event, the fuelwood market can provide useful ways of solving problems of urban timber management that were until recently almost impossible to solve.

The second lesson illustrated by urbanized German silviculture is the unwritten rule that the urban forest must look attractive. It may be highly contrived and artificial, but it must always be scenic with no debris lying around unused or unpiled. The best way I can illustrate this is with an anecdote. A lively and far-traveled German forestry professor, Richard Plochmann, once showed me and two other American forestry professors around a small forest that he managed as a consultant. Each conifer plantation had a strip of hardwoods planted around it, and Plochmann casually referred to these as 'protection strips.' Being inquisitive and analytical, we badgered him about the protective functions and got enough vague or elusive answers that we dropped the matter. Later we came to such a strip, where he had caused two cherry trees to be planted 'because the local people liked the blossoms in the spring.' At that point I asked, 'So, is the protection strip to protect you from criticism?' He laughed and acknowledged that this was indeed the purpose. Of course, the German people have long since accepted the idea that there are so many of them that all land is used—carefully to be sure, but right up to the limit.

Germany is an old country, but Australia certainly is not. Most Australians live in such massive conurbations as Sydney and Melbourne. Some of the areas of highly productive forests are close to these same cities. Most forests are in crown, or actually state, ownership. The Forest Commission of the State of Victoria has been especially aggressive about developing full multiple-use management in a critical area of suburbanized hills called the Dandynongs close to Melbourne. They did this even though they had no charter for anything but timber production. The purpose was to defend their turf from another state agency concerned with wildlife and recreation. It is not quite a case of the forest being trampled by picnickers on weekends and loggers on weekdays, but it comes close.

There is no point in retreating from urban influences and demands. Our purpose instead should be to perceive the nature of the demands and manage forests to meet them. The same silvicultural principles that can be used for growing framing timbers can also be applied to developing attractive landscaping around a factory. In all cases the purpose is to pursue the objectives of the ownership.

Given that this country has a large amount of forest land, it may not always be wise to try to defend every acre of it against encroachment. For example, in southern New England it would be far better to encourage residential, industrial, and commercial development on the plentiful rocky land than to squander the very limited supply of good agricultural land on such uses. We would be better off to work for land use policies that do not waste land of any kind or in any way. Among other things, we are at the end of more than a century of contracting use of land for agriculture, and must, therefore, try to anticipate which lands will be taken from forestry and put back into agriculture. One idea followed in my locality has to do with the most recently abandoned agricultural land, which is either so nearly stone-free that it may be the first to revert to agriculture or so accessible as to be the most suitable for residential or similar use. Therefore, these are not places to plant trees for timber production. Christmas tree production is generally recommended because of the much shorter rotations.

Nevertheless, it is very difficult to manage forests intelligently if the future of land use is always uncertain. There is really nothing that comes closer to defeating real forest management in urbanized regions than this. The general attitude is often 'Next year we will sell it all for a housing development, so why think about growing trees.' I have heard this story for more than thirty years in southern New England. The irony is that next year often doesn't come. The locality now has roughly 17 percent of its land surface in agriculture and about 15 percent in residences, commerce, roads, and industry. The remaining two-thirds is covered with trees. During most of the three decades of which I speak, there were major increases in population. Especially during the 1950s, there was galloping urban sprawl. Half the agricultural land, and more than half of the most prosperous kind, was shifted to allegedly 'higher use.' Some of the poorer kind reverted to forest. Most owners acted as if all the forests were going to be cleared to make way for split-level houses. They scarcely noticed that the total amount of forest land actually increased during this epoch. The whole situation became almost stable during the most recent decade.

In this and some other urbanized regions there are going to be forests, perhaps large areas of forests. This is not necessarily because there is a deliberate plan to have them, wise though that might be. In some cases, it is merely because there is no other demand for the land. Whether the existence of forest land is deliberate or accidental, its management will, in the long run, be better if all concerned know which land is likely to stay in forest for a long time or even indefinitely. Under a passive approach, an assessment of the physical characteristics and accessibility of land will often enable one to predict which tracts of land will almost surely remain under forest cover. With a more active approach, planning and zoning activity may come into play to dictate—subject to appeal—where forests are going to be. Whichever approach is followed, there are some aspects of urban forestry that are bound to limp if future land use is left uncertain.

One need not look far around urbanized areas to detect which land lies idle and drifts like a derelict on the seas of uncertainty. The worst cases are the areas where someone sold all the timber and topsoil before the enterprise went bankrupt.

In many cases, anticipated changes in land use either do not take place or are a long time in coming. For more than thirty years I have presided over a Yale-owned forest on the edge of a small city in New Hampshire. For almost all that time there has been a plan to relocate a state highway over part of the forest. I have thinned the stand three times with a view to developing some pines to sell to the state at a high price. For twenty-eight years I have been repeatedly told that the road will be built 'in five years.' The pines are getting downright overmature.

A young forester of my acquaintance has the appropriate attitude about such matters. He manages a forested municipal watershed which was once threatened with a fevered scheme to sell most of the land for other purposes. He took the position that he would believe the land would be sold only when the ink on the deed was dry. Until then he would continue long-term forest management for water and sawlogs. The sellout didn't happen and his program of forestry is still on course.

THE SCOPE OF URBAN FORESTRY

Urban and suburban forestry is not exclusively a public-sector function. A growing number of private consulting foresters are making a living at it, although perhaps more in the suburban than the truly urban areas. One firm with which I am acquainted has done quite well over many years with such activities as 'forest-scaping,' growing Christmas trees for people on their land, keeping boundaries marked, building fishponds, clearing roads and trails, doing fuelwood thinning, and making timber sales. The owners pay well for the service. The forester-manager told me a while ago of an owner who asked him how far $1,000 per acre would go in making a nondescript red maple swamp look more tidy. The job was done for much less than that, although it was necessary to wait for the ice to freeze solidly to expedite the stand improvement cutting.

Most people, urban or otherwise, tend to regard as wasteland any tracts that are not obviously being used or taken care of. One of the best ways to view urban forestry is to note that the number of potential uses of forest land increases as one approaches the city. The ways increase so considerably that it is pointless to imagine that there is one ideal pattern or mode of use. Just as out in the rural forest, the advocates of each kind of use will regard every acre diverted from their purposes as a desecration. The zoning of uses and ranking of use priorities, tract by tract, is probably even more important in urban forests than it is in more remote public forests also dedicated to multiple use.

One sad phenomenon associated with noble intentions and urban influences is that of the quasi-public conservation area presided over by a committee of dedicated amateurs. They typically and collectively want every use or nonuse piled on each acre. Usually they fail to resolve their disagreements so that nothing happens and onlookers conclude that the area is in some limbo of futility. In this case, as in so many others, there is an important role for foresters in helping landowners sort out their objectives.

The closest thing to a universal benefit that is sought from urban forestry is beauty. This is a commodity long known to exist chiefly in the eye of the beholder. It can often be quite artificial. During the period that the forestry profession was desperately trying to convince the forest industry that forestry was a paying proposition, it developed a trained incapacity to recognize beauty and other attributes not readily translated into merchantable timber. At the other extreme, beauty is not necessarily what an ecologist would recognize as natural. For most people one important kind of forest beauty would be typified by a Douglas-fir plantation with a lawn of some kind of Eurasian grass beneath it. However, that isn't the only kind.

It is not difficult to list some things that are not beauty. Unused logging debris leads the list, closely followed by the many slovenly practices all too commonly associated with logging. It is providential that the fuelwood market has reappeared to help deal with logging debris. Attitudes about shrubby understories are variable, although a complete stratum of this kind of vegetation is almost always regarded an an unkempt mess of brush. Various scatterings of shrubs can, on the other hand, be very attractive. It helps to think in terms of developing focal points, such as large trees, rocks, or flowering shrubs, in arrangements that people will look at. Conversely, since people tend to stick to roads and paths, it is not difficult to develop screens or other ways of hiding the unattractive things.

Another major item of urban forestry is microclimatic and environmental modification, chiefly through the provision of shade. It has been pointed out in this connection that if the deciduous tree had not existed in nature, engineers would have had to create it. However, it takes knowledge, patience, and hard work to maintain this kind of shade where it is needed most.

The most important places for making recreational and educational use of forests are as close as possible to cities and not off in some remote wilderness area. It is sad to observe how many well-supported nature-education centers languish off in rural areas. The most successful one that I know about is a patch of woods next to a high crime public-housing area; it is run by an enthusiastic naturalist who has a way with kids, including the kind who might otherwise be in training for the state prison. So it is also with sylvan recreation parks and swimming areas: we can never have enough close to the cities.

Forestry also plays a crucial urban role in the maintenance of forested watersheds. In a locality such as southern New England, these constitute almost a twentieth of the land surface. These water supply systems are administered by a bewildering mixture of semipublicly owned utilities, municipalities, and regional public authorities. The land management personnel may be farmers, engineers, foresters, or politicians. The areas that are managed best and most economically are usually those where some forester is in charge of the lands. In fact, these tracts have some of the best forest management in the region.

As far as wildlife management is concerned, fishing and passive observation of wildlife thrive rather well in the urban forests, but it is difficult to make hunting compatible with the situation. In fact, it is an environment in which it is easy for herbivores of all sorts to get out of control. It is probable that domestic dogs and cats exert more control over urban mammal populations than anything done directly by people. This would seem a rich but uncultivated field for wildlife management research.

Timber management should not be regarded as off limits in urban forestry, even though it will almost always be a secondary, tertiary, or incidental objective. There is no way that forest vegetation can be manipulated or regenerated for any management purpose without killing some trees to make vacant growing space. The most economical way to kill trees is to cut them down and use their substance. If wood were useless, the treatments required for the conduct of urban silviculture would be even more costly than they are. There are plenty of tracts in urban areas where timber management can be made compatible with the uses of higher priority. Furthermore, under the American system, whether the area be rural or urban, timber is almost the only source of direct revenue for the landowner. However, this point must be kept in its proper perspective.

One of the biggest impediments to using privately owned urban or suburban forests for timber production is the doctrinaire view that this isn't feasible unless the timber revenues will pay all of the cost of owning the land. Foresters can be even more afflicted with this misconception than landowners. There may be all kinds of reasons, mostly not regarded as rational by economists, why such lands are kept as forests regardless of costs, taxes, or opportunities forgone. If they are going to be kept as forests for any significant period, there is no fundamental reason why they should not be managed for timber. It helps pay the high costs. For some owners it may be a hobby like growing roses. For others it may be a source of firewood and exercise. There are cases in which society would do well to subsidize timber production on certain urban lands that become problems if not kept in forest cover.

Urban forestry is not necessarily a jungle bristling with insoluble problems. Actually it is an opportunity for the forestry profession to exhibit its prowess where people can see it. The results will be judged by how well they work and not by any long-faced calculations of present net worth in dollars. It is an arena in which foresters can win, but they will have to alter some habits of thought and ways of doing things if they are to prevail. If they can win, what better showcase could they have to convince people of the wisdom of sound silviculture regardless of where it is practiced?

14

WILDLIFE CONSERVATION AT THE URBAN/FOREST INTERFACE

Richard D. Taber

European settlement of North America has been characterized first by the establishment of multitudes of small family farms, followed by a consolidation of farms into larger units and an increase in the size of cities. Over the years the urban proportion of the human population has steadily grown, and the rural proportion declined.

Now a new pattern is emerging. There is a movement of educated, affluent urbanites into the suburban and rural regions. This is a development that is potentially favorable to wildlife conservation. The reasons for this, and some of the things that might impede or promote the realization of this potential, are the theme of this chapter.

The most productive lands in North America, the lands that supported most of the enormous populations of wildlife that European settlers found on this continent, are now in private hands. The private landowner has usually managed his land for the production of crops. Agriculture, in the broad sense, has become ever more intensive and productive of crops that can be sold.

Wildlife is not a crop that can be sold in North America, although it is a common item of commerce elsewhere. With the establishment of the United States, wildlife became the common property of the citizens of each state. This public ownership became private when some private party gained possession of a wild animal. Private individuals, then and now, operate on two planes: they are business persons, making decisions on the basis of costs and benefits, and they are household persons, making decisions on the basis of their own and their family's desires.

The pre-1900 era of wildlife acquisition was pursued by both business persons ('market hunters') and household persons ('sportsmen'). Under the combined pressures of market hunters and sportsmen, wildlife had declined by the late 1800s. Eventually the sportsmen organized, promoted the establishment of state wildlife agencies, and, perceiving wildlife commerce as a threat to their desired recovery of huntable wildlife populations, worked successfully for the prohibition of wildlife commerce.

The annual license fees for hunting supported the state wildlife agency, which in turn promoted the sportsmen's interests. One such interest was access to private lands in pursuit of game. This access is made easy, in most states (Texas is an exception), by trespass laws that do not provide the private landowner with automatic control of access by others to his land. The crop of game animals produced on private cropland is harvested by the sportsmen, with no benefit to the land manager. Since wildlife production is not one of his managerial objectives, wildlife habitat and production on private croplands have declined radically in North America despite all sorts of efforts to slow or reverse the trend.

Now private croplands are being subdivided into 5-, 2-, or 1-acre lots for the new rural residents. The acre that yesterday had a zero or negative value for wildlife, because wildlife did not constitute a commercial crop, tomorrow will be some new rural resident's backyard, where wildlife will have a high amenity value. This is a fact that poses a challenge to traditional thinking about wildlife by utilitarian Americans, and a potential new opportunity for wildlife conservation.

Traditional American wildlife thinking has been game oriented. Leopold's classic book on wildlife conservation in North America (1933) was entitled *Game Management*. State departments responsible for wildlife have been called Game Departments. They have been supported by hunting license sales and taxes on sporting arms and ammunition used in the pursuit of game.

A broader view of wildlife values was not completely lacking, as the protection of so many species by the Migratory Bird Treaty of 1918 shows, but at the state level the emphasis on game birds and mammals—only about one-tenth of all wildlife species—was, and is, built into the conservation system.

Outside that system, the American public was becoming more and more concerned about environmental quality. This concern was crystallized, in 1962, by Rachel Carson's *Silent Spring*, and through the 1960s and early 1970s there was much new legislation aimed at protecting endangered wildlife environments.

From the vantage point of 1982, we can see that the greatest impact of environmental concern on land management has been on public lands. These have traditionally been managed by crop-oriented professionals for economic return, but the public has consistently viewed public lands in a household not a business sense. The changes in public land management policies forced by this public attitude promise a brighter future for wildlife habitat and wildlife conservation on public land.

On private lands, as shown above, economic cropping and wildlife conservation are generally not compatible in North America. But the shift in land use from the business use to the household use with the shift in human population from the city to the country has opened new prospects for wildlife. Among the factors that favor wildlife increase on these rural residential lands are the interest in wildlife, the willingness to pay, the level of education, and the appreciation of environmental quality by the new rural immigrants.

Recognition of the wildlife conservation opportunity on these new rural residential lands has naturally lagged behind the actual demographic shift itself. In fact, subdivision of croplands is often viewed with alarm in wildlife conservation circles. It increases the work of state conservation agencies by increasing the involvement of wildlife in nuisance situations (five hundred raccoons are trapped and translocated from Seattle's suburbs annually), and by increasing hydraulic permits, environmental impact statements, zoning variances, and the like, which must all be reviewed. And every farm subdivided reduces the hunting opportunity for the sportsmen from whom state wildlife agencies draw their principal financial support.

The opportunities provided by the proliferation of rural residential lands for wildlife conservation are only beginning to be recognized and, in a piecemeal fashion, responded to. The western Washington example is illustrative. The three most urbanized counties—Snohomish, King, and Pierce—have more residents outside of incorporated areas than in them. The increasing proportion of exurbanites among them tend to regard wildlife not as a source of food but as an object of interest, even when there are deer in the garden or skunks under the porch. Recently, when the extension naturalist of Snohomish County offered a short course on 'Ponds for Wildlife,' 250 citizens enrolled. Under his tutelage, Boy Scouts and 4-H groups build and place nest-boxes for wood ducks and platforms for Canada geese. The interest is there.

A program to turn this situation to best account for wildlife conservation might well be fully developed from such small beginnings by focusing on the attainment of the three general objectives of wildlife conservation: conservation of genetic diversity, enhancement of diversity and abundance, and alleviation of pest situations. In pursuing each of these objectives it would be recognized that the movements of wild species carry them daily or seasonally over a much larger landscape than that of a single property owner, so that there must be local and regional coordination, at least, in the development of wildlife conservation plans. This constitutes an opportunity for cooperation between agencies and citizens' groups concerned with wildlife conservation and those concerned with land use planning.

Conservation of genetic diversity can be achieved through the restoration of endangered and threatened species to secure population levels. The U.S. Fish and Wildlife Service identifies such species on international and national levels; the state conservation agencies add species of state concern; and informed citizens' groups such as the Audubon Society identify populations that appear to be waning. From such sources a list of species of special conservational concern could be developed for any particular locality, and requisites for their encouragement detailed.

Local diversity and abundance of most wildlife species are controlled by opportunities for colonization and by quality of habitat. The former can be enhanced by regional and community action favoring contiguous travel ways. The immediate habitat on residential property can be improved in a great many ways. Making current information available to landowners, filling gaps in ecological knowledge, and experimenting with habitat enhancement for species in need will all be useful.

Alleviation of pest situations is now approached, in large part, by removal of offending individuals. There is much opportunity to minimize damage by protecting the things now being damaged, by substituting things less prone to damage, and by diverting the wildlife species now involved in the damage.

A comprehensive program of wildlife conservation that encompasses the variety of ecological and social situations from the urban through the rural regions, developed on the principles sketched above, and fueled in large part by the financial and political power of the new rural immigrants, would be something new for North America, and something of value to pass on to future generations.

15

SOIL MANAGEMENT IN URBAN FORESTRY

Byron R. Thomas

My discussion will be related to David Smith's statement that one of the most helpful kinds of information that foresters can bring to the city from the wildwood is the knowledge of soil conservation. Awareness of urban soil management problems could help prevent them from occurring.

Soil, as conceived by the Soil Survey Staff (1975) in *Soil Taxonomy*, is the collection of natural bodies on the earth's surface, in places modified or even made by man of earthy materials, containing living matter and supporting or capable of supporting plants out of doors. This concept allows us to consider most areas in the urban environment that support trees as having soils. It also forces us to realize that once a soil is dug up and put in a pot or container, it becomes a 'nonsoil' even though it is a medium for supporting plants.

The important consideration is that a medium—soil or nonsoil—for supporting plants may be damaged to the extent that it does not fulfill its purpose to produce benefits to its potential.

Craul and Klein (1978) studied some characteristics of urban soil properties. They reported that urban soils generally have high pH values and salt contents, moderate water-holding capacity, and low organic matter content when compared with nonurban soils. Many urban soil profiles showed incipient mottling, which indicated aeration problems. All these factors adversely affect plant growth.

Gray (1978) recognized that urban soils are drastically altered. Building sites, streets, parking lots, and so forth, actually destroy the soils. He found that adjacent areas where trees may be planted are often subjected to the same processes.

One of the most common afflictions that man imparts on the soil, urban or nonurban, is compaction. Chancellor (1977) described soil compaction:

> The total volume of a soil is made up of the volume of soil mineral grains and the volume of the pores between the grains. The pore volume is usually partially filled with water, with air occupying the balance. Soil is compacted when the proportion of total pore volume (or air-filled pore volume) to total soil volume is inadequate for maximum crop growth or efficient field management. When crop growth appears inadequate and the soil seems hard, soil compaction is often implicated—quantitative measurements arc important, however, for accurate problem diagnosis. If soil compaction is not at fault, other potential problems can be isolated.
>
> Soil compaction can be considered in terms of soil porosity and density, water infiltration, or penetrometer resistance.

Vomocil (1979) colorfully described the effects of soil compaction on plant growth:

Soil is a 3-phase system—solids, liquids, and gases. Each independently, *and their* interactions, and their relative motions determine the suitability of soil as a root medium for plant growth. A soil is a leaky tank filled with sponges, springs, dashpots, pipes, orifices, valves and other fluid control devices. The health and welfare of plants is a fluid mechanics problem and the soil is the medium where the currents, counter currents and response currents occur.

Compaction breaks springs, jams dashpots, smashes sponges, bends and kinks pipes, clogs orifices, gums up valves, and in general raises hell with soil as a fluid conductor and as a medium for plant growth.

1. Compaction reduces soil aeration, inhibits root system respiration.
2. Compaction modifies water supply to roots—viz., changes hydraulic conductivity.
3. Compaction increases the mechanical impedance of soils to root system; roots are restricted, short, stubby, deformed, and shallow.

Russell's (1979) book treats the interaction of roots with soils at great length. It is recommended for those who wish to obtain a general understanding of the soil-root interaction.

Patterson (1976) reported some very interesting comparisons of bulk density and pore space for various materials. These are summarized in table 1.

Use of off-road vehicles (ORV) has grown tremendously over the last few years. ORV activity occurs in the urban areas as well as open spaces. Uses of ORVs in the urban environment range from legitimate uses to destructive acts of vandalism. McKnelly (1980) reported that in one study area the acreage receiving wheel-to-ground impact increased from 13.6 acres in 1973 to 35.7 acres in 1979. Erosion also increased. Average trail width increased 31 percent, and some damage to tree roots at creek crossings occurred.

Table 1. Average Bulk Density and Pore Space for Various Materials.

Material	Average Bulk Density (g/cc)	Average Pore Space (%)
Ideal soil for area	1.4	40.0
Surface soil from Washington, D.C., Mall	2.13	19.0
Brick	1.75	35.7
Asphalt	2.19	6.8
Concrete	2.26	8.5

Numerous studies show how compaction adversely affects tree growth. Adams and Froehlich (1981) summarized seven studies that reported the relation between seedling height growth and percent increase in soil density (figure 1). Their work shows that seedling height growth decreases as soil density increases. They pointed out that regeneration in compacted areas did not necessarily equate with future growth.

Zisa et al. (1980) reported that when northeastern coniferous species were grown in compacted silt loam and sandy loam, establishment was achieved on a bulk density of 1.8 g/cc in the sandy loam soil and between 1.4 and 1.6 on the silt loam. Root penetration was markedly reduced at a bulk density of 1.6 on the sandy loam and 1.4 on the silt loam.

A study by Wert and Thomas (1981) quantified effects that compacted skid roads had on wood production. Their study was designed to obtain information on the long-term effects skid roads had on Douglas-fir growth and determine the degree and extent of soil compaction thirty-two years after harvest.

They categorized the study area into skid roads, transition zones, and undisturbed sites. Skid roads had 41 percent fewer trees and 75 percent less volume on an acre basis than undisturbed sites. The transition zone (10 m on each side of the skid road) had 17 percent fewer trees and 25 percent less volume than the undisturbed sites.

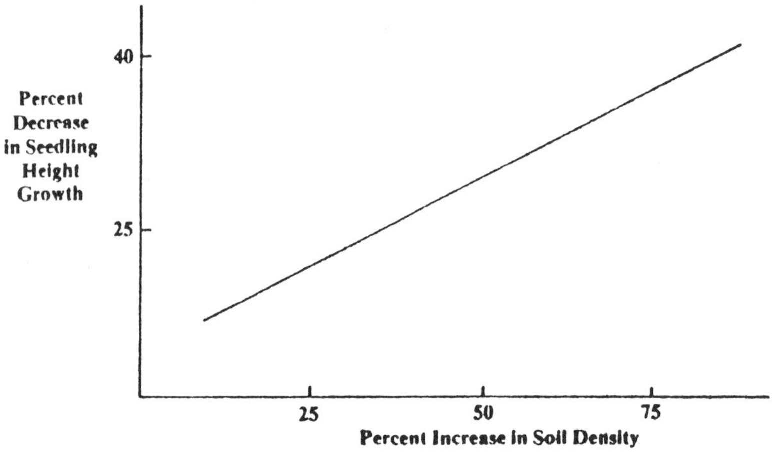

Figure 1. Relationship between increases in forest soil bulk density and the height growth of tree seedlings (based on field studies of a variety of tree species, soil types, and locations).

Volume losses were due to the trees in the skid roads and transition zone being shorter and having smaller diameters (table 2). Skid roads were still heavily compacted below 15 cm after thirty-two years.

Some ailments of urban soils can be treated so their productivity may be maintained at or close to potential. Nutrient deficiencies are undoubtedly the easiest to correct. These can be corrected whether the site is vegetated or bare. Nutrient deficiencies may be detected by soil or foliar analysis and plant symptoms. Local extension services, experiment stations, and consultants may be called on to define the kinds and quantities of nutrients to be applied.

Toxic agents that may become incorporated into the soil are usually more difficult to work with than nutrient deficiencies. The toxic substance must first be identified and then correction may or may not be feasible. A higher level of expertise is usually required to deal with toxicity in urban areas than that needed for nutrient deficiencies, because the toxic substance may also be hazardous.

Amelioration of soils that were badly treated adjacent to construction sites is best accomplished prior to planting. This would include debris removal, tilling to break up soil compaction, and compensating for loss of productivity by incorporating proper soil amendments and nutrients.

Table 2. Effects of Skid Roads on Mean Tree Heights and Diameters. Comparisons of the Variable Means by Student's t-test.

Category	Mean Tree Height (m)	Mean dbh (cm)
Skid roads	10.8**	11.7*
Undisturbed areas	15.4	16.3
Skid roads	10.8**	11.7**
Transition zones	14.1	15.5
Transition zones	14.1	15.5
Undisturbed areas	15.4	16.3

*Significantly different at the 0.05 level.
**Significantly different at the 0.01 level.

Correcting physical impairments while maintaining plant growth is a challenge. For example, what treatment is feasible to correct heavy soil compaction under foot trails in forests? Mechanical tillage is required to break the compacted layer in areas where there is continuous use. Tillage also injures the roots and makes the wounded areas susceptible to diseases. Good planning and project design that address potential problems are probably the best and most economical long-term mitigating measures.

REFERENCES

Adams, P. W., and H. A. Froehlich. 1981. Compaction of forest soils. A Pacific Northwest Extension Publication. PNW 217. Oregon State University Extension Service, Corvallis, Oregon.

Chancellor, W. J. 1977. Compaction of soil by agricultural equipment. Bulletin 1881. Division of Agricultural Sciences, University of California, Davis.

Craul, P. J., and C. J. Klein. 1978. Characterization and interpretation of urban soil properties for tree growth. In: Vol. 2 of Proceedings of the National Urban Forestry Conference, Washington, D.C., November 13-16, p. 286.

Gray, G. W. 1978. Urban forestry. John Wiley and Sons, New York.

McKnelly, P. K. 1980. Turkey Bay off-road vehicle area: Its use and monitoring system. In: R. Andrews and P. Nowak, eds., Off-road vehicle use: A management challenge, pp. 91-99. Sponsored by the Office of Environmental Quality, USDA, School of Natural Resources, University of Michigan, University of Michigan Extension Service.

Patterson, J. C. 1976. Soil compaction and its effects on urban vegetation. In: F. S. Santamour, Jr., H. D. Gerald, and S. Little, Jr., eds., Better Trees for Metropolitan Landscapes Symposium Proceedings, pp. 91-102. U.S. Forest Service NE-22.

Russell, R. S. 1979. Plant root systems, their function and interaction with the soil. McGraw-Hill Book Company (UK) Limited, London.
Soil Survey Staff. 1975. Soil taxonomy. Agricultural Handbook No. 436. U.S. Government Printing Office, Washington, D.C.
Vomocil, J. A. 1979. Soil compaction: Processes and measurements, workshop handout. Sponsored by Forestry Intensified Research Program, Oregon State University, Medford, Oregon, September 18-20.
Wert, S., and B. R. Thomas. 1981. Effects of skid roads on diameter, height, and volume growth in Douglas-fir. Soil Sci. Soc. Am. J. 45:629-632.
Zisa, R. P., H. G. Halverson, and B. B. Stout. 1980. Establishment of early conifers on compact soils in urban areas. Northeast For. Exp. Sta., Broomall, Pennsylvania. U.S. Forest Service NE-451.

PART V

SOCIAL AND PSYCHOLOGICAL DIMENSIONS

16

IMPLICATIONS OF CONTEMPORARY COMMUNITY ORGANIZATION AND SOCIAL VALUES FOR FOREST MANAGEMENT ON THE RESIDENTIAL/WILD LAND INTERFACE

Robert G. Lee

Contemporary American foresters are rediscovering social issues once faced by their European progenitors. People living in or near forests influence the purposes for which forests are to be managed as well as the methods for fulfilling those purposes. Nothing was more central to the development of European forestry practices than the competing uses that increasing numbers of local residents made of forest lands. Today's foresters are struggling with that same problem, since they must accommodate an increasing number of competing uses on a shrinking land base. Even though the specific character of forest uses may differ from those of medieval Europe, the social problems of accommodating a proliferation of uses for forest lands, particularly on those lands near the fringe of residential development, have led people to talk about the 'urban/forest interface.'

I prefer the term 'residential/wild land interface' to 'urban/forest interface.' The word 'residence' denotes the 'act or fact of dwelling in a place for some time'; 'urban,' by contrast, offers a far less precise meaning: 'of, relating to, characteristic of, or constituting a city' (Webster's). Most of the social problems faced by foresters are not in cities and do not involve only people living in cities. People make use of forested lands for second homes, industrial development, and temporary squatting as well as permanent residence. Problems arising from these uses involve proliferating claims to forest land in nonmetropolitan areas. Moreover, a focus on residential use instead of an urban setting more clearly shows how forestry itself originated on an 'interface' between forest uses by local residents and the claims of the European nobility. Attention to residential uses also serves to explain how transportation systems, communications, and services make the differences between urban and nonurban living increasingly difficult to detect. The term 'wild land' is more appropriate than 'forest,' since competing claims are made to brushlands and grasslands as well as to forest lands.

My discussion of the residential/wild land interface will begin with a more comprehensive delineation of its attributes. Next I will discuss how forestry practices emerged as a product of a residential interface in medieval Europe. The discussion of historical antecedents will be followed by descriptions of how advanced industrial society has influenced social values, patterns, and ecological linkages relevant to forest management. I will conclude with comments on the implications for forest policy of emerging values, settlement patterns, and ecological linkages.

DELINEATING THE INTERFACE

Many attributes have been abstracted from the circumstances observed when residential development occurs in or adjacent to wild lands. The 'urban/forest interface' has been described in geographic terms as a 'location where urban/suburban development and forest-resource management go on simultaneously in the same area' (Vaux 1982, p. 37). Such a definition draws attention to the fact that people's homes or places of work occupy land areas that are also dedicated to the maintenance of forests.

Vaux assigned a broader meaning to the urban/forest interface when he defined it as 'a *political* phenomenon, not a geographic one. The interface—or interaction, or confrontation—is between different political forces, each of which has a somewhat different set of perceptions about the value of forests to society.' This definition directs attention to the political groups that compete either for the use of forests or for influence over how forests are to be used by others. However, according to this definition, *all* forest policy issues would be interface issues, since any controversy over the use of forests involves 'differences in view over the social values to be emphasized in forest policy.'

The interface can also be characterized as an ecological condition in which human activities and artifacts interact with natural objects in wild land ecosystems. This is illustrated by the residential/wild land interface fire hazard problem: the interface consists of the rapid transfer of heat from wild land fuels to fuels found in residential structures, or vice versa (Lee 1982). Another significant problem involving undesirable ecological linkages is the adverse impact of household pets, particularly dogs and cats, on natural wildlife populations and domestic livestock.

An interface can be more accurately delineated by distinguishing between its geographic, social, and ecological components. Greater clarity in the use of these terms can be achieved by adopting the logic of set theory: union, intersection, and set inclusion (see figure 1). The union of all three components represents a situation in which political groups assign competing values to particular ecological linkages between human activities and a natural ecosystem in an area where residential development is located within or near a forest. Such a near union of sets is illustrated by local residents who are in conflict with wildlife biologists and hunters over the disruptions of migratory deer populations caused by dogs that residents feel should be 'left free to roam in the open country.' Alternatively, both the ecological and geographic components could be subsets of the political components when the dog-wildlife conflict is viewed simply as a part of a national political conflict over the extent to which the wild land environment should be disrupted by agents such as logging, biocides, or domestic animals. An intersection of components is illustrated by local residents who are in conflict with forest managers because they believe spraying herbicides near their homes will make their families sick and, in addition, feel that biocides hurt all forest ecosystems and people who live near the forest. Most interface issues are extremely complex, since they involve intersections in varying degrees between geographical, political, and ecological components.

FORESTRY AS A PRODUCT OF THE INTERFACE

Interface problems are by no means new, even though most foresters tend to view them that way. Attempts to manage a residential interface motivated the early development of forestry practices. European forestry policies and practices originated at least nine centuries ago as methods for protecting the hunting privileges of the nobility. In England, William the Conqueror and his successors established vast royal forests, within which the forest law forbade the killing of the king's deer (Young 1979). Subsequent elaborations of these early laws restricted the uses

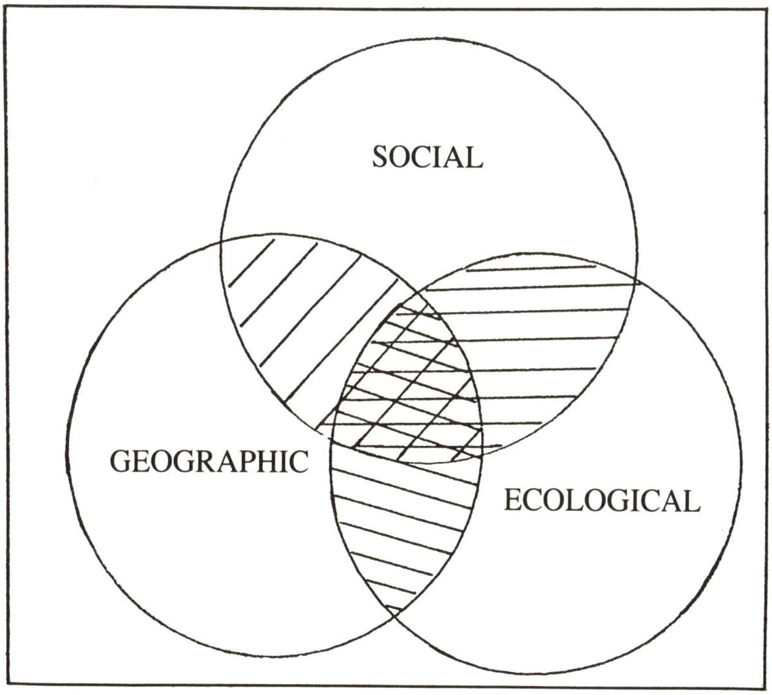

Figure 1. Delineating the interface.

made of these forests by local residents, most of whom were subsistence farmers living on feudal estates. The relation of these forests to the lives of medieval people can be described in terms of the geographic, ecological, and social components of the interface existing at that time.

An understanding of this historic interface is particularly important for appreciating how the political and professional role of the forester was shaped by history. Many of the foresters' problems with learning to manage an interface today can be traced to the trained incapacities, or *idée fixe,* acquired during the formative stage of the profession. Let us examine how foresters acquired a distinct social role.

Throughout European history the vast majority of people were subsistence farmers living in the countryside. An expanding population was accommodated by clearing the forests and establishing agricultural settlements in their place. Most of the lowland areas of Europe were cleared of forest by the 19th century. The remains of natural forests persisted on uplands where the soils were unsuitable for growing agricultural crops. With the arrival of the industrial revolution in the 18th century, upland forests were harvested rapidly for fuel and building materials. The profession of forestry was developed to manage these remaining forests for continuous wood production, since the local communities and industries, particularly metallurgy, required large amounts of fuel and construction materials. Yet the cultural roots of forestry reach back to feudal times.

European farmers had always treated forested areas as residual wastelands that were held in common by the local residents. People established a number of ecological linkages with these wastelands, despite their relatively low value. Forests provided wood for fuel, implements, and building materials, honey for wax and sweetener, peat for fuel, acorns for hog food, grasses and herbs for domestic livestock, litter for fertilizer, and wild animals for food. In addition, forests provided an outlet for people displaced by social and economic change. Landless peasants would seek the 'frontier' and clear its forests in order to grow food. Residual lands covered by forests were sufficient for this purpose until the 18th century, when rapid population growth, consolidation of agricultural ownerships, and the emergence of the industrial order combined to squeeze peasants out of agriculture and into the cities. The poor laws superseded forests as a means for supporting the landless. In sum, the removal of Europe's forests facilitated agricultural and industrial development.

Political support for the protection and management of forests emerged first with the hunting interests of the nobility. The first foresters were medieval game wardens, retained by the nobility to protect both the game and the habitat on which it depended. Local subsistence farmers and landless peasants were the main threats to forests and the game they supported. Thus the first forestry practices were developed by people whose interests were directly opposed to the traditional forest uses of subsistence farmers.

The political posture of foresters who developed the profession in the 18th century did not differ substantially from that of their medieval predecessors. Most common forest lands had, with the collapse of the feudal order, been appropriated by the nobility, and were subsequently retained in private ownership or transferred to the state. Thus foresters worked for the state, large private owners, or the remaining nobility. In all three cases, local residents were seen as threats to the integrity of the forest, and the needs of local people were accommodated only when necessary to prevent revolt or uncontrolled use (Fernow 1913).

A strong tradition of protecting forests from destruction by local people emerged from these early European experiences. This tradition has persisted in most countries that have adopted European forestry policies and practices.

The American forestry profession owes its origin to the German forestry tradition. German foresters enjoyed a position of high social standing that aligned them politically with the interests of large landowners and the nobility. American foresters did not enjoy similar social standing when they adopted Germanic traditions eighty years ago. It required the political genius of Gifford Pinchot to link forestry to the agricultural and ranching interests in the West and the intellectually oriented conservation interests of people living in the East. These strategic political ties were formed so that foresters could gain the support needed to manage 'for the greatest good for the greatest number in the long run.' Foresters found it difficult, if not impossible, to respond to the interests of farmers who had cleared forests in order to establish agriculture in the East, South, and Midwest. Also, foresters had an ambivalent relationship with the wood industries that had cleared the forests in order to provide products essential for building the nation's railroads, manufacturing centers, and cities. Foresters were carried along by the politics of a national conservation crusade that was dedicated to saving forests from further destruction. More productive uses for forested lands and more efficient uses for wood did not receive much attention. Like their European progenitors, American foresters pursued an extremely conservative approach to forest land use.

Why does the contemporary forester find the residential interface so troubling? What trained incapacity limits the ability of the forester to address this problem? Some critics claim that foresters are more concerned with the forest as a biological entity than with meeting the needs of people. This is a specious accusation. We have learned that, by tradition, foresters are concerned with *some* people—the large landowners and the abstract 'people' collectively represented by the state—but are less concerned with local residents and the landless. In short, foresters have served effectively as agents for the interests of large public and private landowners, but are still struggling to learn how to serve other interests that make claims to the use of forested land. Foresters cannot contribute effectively to solving the 'problem' without confronting their own incapacity to manage lands in response to the interests of a wide variety of claimants. A closer examination of the residential/wild land interface problem will reveal how foresters might respond to the proliferation of competing claimants.

Discussion of the residential/wild land interface becomes most relevant to foresters when its components are specified in enough detail to show how concrete actions can be taken. Although we talk about *the* interface as if it were a singular problem, there are actually innumerable interface situations. Management of specific interface conditions requires clear delineation of their components. Social, geographic, and ecological components will be discussed in the context of an advanced industrial society. The political component will be addressed by describing some of the changes in social values associated with an increasing number of competing claims to the use of forested lands. The relation between these components will be explored in a discussion of forest management implications.

INTERFACE PROBLEMS IN ADVANCED INDUSTRIAL SOCIETY

Social, geographic, and ecological components of the contemporary residential/wild land interface are shaped by conditions in modern society. Modern technology, economic organization, and social values are especially important. Sociologists have used the term 'advanced industrial society' to denote a society increasingly dependent on rapid advances in communications and transportation technology, growth in the service sector of the economy and *relative* decline in the importance of primary resource extractive industries, and growth in both public goods, such as environmental quality, and public interest groups organized to promote or defend public goods (Bradshaw and Blakely 1979). In particular, these social and economic conditions are reflected in the changing occupational structure of the economy.

Figure 2 shows the relative distribution of occupations by the primary, secondary, and tertiary sectors of the economy for the United States and Washington State from 1910 through 1979. Employment in the primary sector (agriculture and extractive industries) has declined from roughly a third of all employment to less than 5 percent in seventy years. Employment in the secondary sector (manufacturing and construction) has remained relatively stable at about 30 percent of total employment, while employment in the tertiary sector (trade, finance, insurance and real estate, transportation, communications, utilities, public administration, and the service industries) has in seventy years grown from about 40 percent to almost 70 percent of total employment. The state of Washington has shown greater reliance on employment in the tertiary sector and lesser reliance on the primary sector than the nation as a whole.

Some Distinguishing Characteristics of Advanced Industrial Society

These trends have profound implications for forestry, both indirectly, by stimulating population redistribution and change in tastes and values, and directly, by altering the political relationships between forest-based industries and an electorate that is increasingly better educated

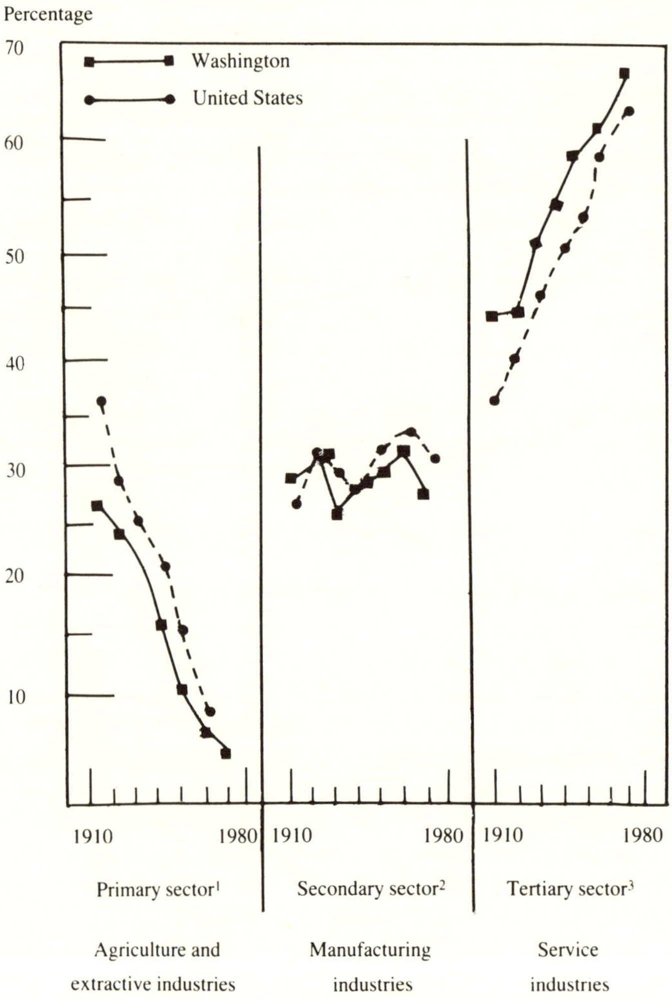

Figure 2. Trends in Washington State extractive and nonextractive economic sectors.

1. Includes agriculture, forestry, fishing, and mining industries.
2. Includes manufacturing and construction industries (including logging and wood manufacturing).
3. Includes trade, finance, insurance and real estate, transportation, communications, utilities, public administration, and the service industries.

Source: Calculated from U.S. Bureau of the Census, *Census of Population, Detailed Characteristics* (Washington, D.C., 1910, 1920, 1930, 1940, 1950, 1960, 1970; 1980 census data are unavailable). Data are generally comparable over time, although some differences exist because of the age groups included in employment counts: through 1940 children over 10 were included, by 1950 only workers over 15 were included, and by 1960 the age group included 16 and over.

and concerned with environmental amenities. Following is a summary of some principal distinguishing characteristics of advanced industrial society.

Growth of the service sector has been accompanied by a higher average level of education and greater specialization in skill-related occupations. Computer technology illustrates this trend. This change is associated with growth in the importance of technological innovation to economic growth. However, growth in the service sector at the expense of the primary sector has resulted in a shift of employment from occupations closely related to land and its associated resources to tertiary occupations far removed from basic production processes. This is important for foresters and planners to comprehend, since a high proportion of the work force lacks an appreciation for the problems of deriving resources from land. The political implications of this change for land-based professions and industries is already being felt in the emergence of environmental quality controls.

Advanced industrial society is also accompanied by the proliferation of interest groups and growth in their importance as a means of political expression. Citizens mobilize and form interest groups in an attempt to influence both governmental institutions and private industries. Many of these groups take the place of traditional local communities as a form of social organization that stands between the individual and the larger institutions of society. As such, they embody a multitude of values, not the least of which are the values of a clean and livable environment.

Advanced industrial organization relies on the growth of a social and economic support system necessary for the functioning of a service-based economy. Growth of a high quality highway system, regional airports, local television programming, and advanced telecommunication services have afforded people opportunities to move to locations where work in service industries can be combined with an attractive residential environment. Personal freedom of mobility and relatively unrestricted land markets have facilitated the rapid growth of communities in rural locations as well as on the urban fringe.

An important consequence of these technologically induced demographic changes is a decline in the importance of both occupational and geographic communities. People no longer identify as closely with their work as a means of defining their identity and place in society, and traditional rural communities are no longer as homogeneous and cohesive. These changes are important for the forester to note, because there has been a substantial decline in the difference between 'urban' and 'rural' societies. It is no longer safe to generalize about the distinctive beliefs and attitudes of urban and rural residents, because rural communities increasingly consist of recent immigrants from urban areas.

Consequences of Advanced Industrial Society for Social Values

Both the number and intensity of demands on wild-land use have been increasing. Political conflict centers most often on the protection of natural ecosystems and restriction of the technologies applied to their management. Although wilderness protection has captured political attention at the national level, the protection of small tracts of wild land adjacent to residential areas is increasingly a concern of local citizens. In the state of Washington such concern has resulted in political conflict between local planning agencies and forest landowners. The Washington State Department of Natural Resources is severely affected by such conflicts, since it administers thousands of small parcels of forested school trust lands in regions of the state that are increasingly occupied by residential development. However, there are many other types of political conflict, some of which do not represent such a clear intersection of geographic and political components of the residential interface.

Social conflict over the use of wild lands has several sources. All conflict is associated with the competing values people assign to land and the resources it provides. Some values are highly localized, such as the sentiments that local residents acquire for adjacent wild lands. Other values are diffuse. The emergence of advanced industrial society has been associated with several diffuse values—open space, natural scenery, water quality, air quality, and outdoor recreation. All of these values take the form of public goods and are thus usually provided by public agencies. However, residents who migrate to a location to enjoy the amenities of a natural environment may seek to protect the environment from further development by restricting the uses to which private lands are allocated.

Local movements to limit growth are motivated by the desire to maintain the residential environment in the state encountered by recent migrants. I will refer to this phenomenon as the 'last settler syndrome' (Nielson et al. 1977) since it reflects the values assigned to the environment by residents who have just arrived. In other words, people move into an area because it offers natural amenities and the values associated with relatively low-density living; then they try to limit growth in order to maintain the environment as they found it. Like Daniel Boone, they are dissatisfied when they can see the neighbor's smoke, but unlike Boone, they use political tactics in an attempt to keep new neighbors from moving in. Each wants to be the last settler.

The last settler syndrome is also expressed in the shift of values migrants undergo. There is affirmation of freedom of mobility and uncontained land markets by aspiring migrants seeking a home on the edge of the forest. Yet these same individuals promote interference with land markets and exclusive community standards to defend the 'rural' or 'uncluttered' character of the community they succeed in joining.

Those who favor growth limits also find themselves in conflict with landowners and developers who claim the right to use their lands for residential and industrial purposes. Forest landowners in western Washington are in an ambivalent position. They seek to restrict development on lands from which they may derive substantial returns from timber harvesting, but favor the development of those same lands when they will gain a greater return from a conversion in land use. Recent migrants are in turn ambivalent toward forest landowners. They tend to want forested lands to remain dedicated to forest resource management, but object to timber harvesting and silvicultural treatments such as herbicide spraying.

Social conflicts also emerge between recent migrants and longer-term residents. Migrants to nonmetropolitan areas tend to bring modern ways with them, thereby imposing the values of an advanced industrial society on people who may still cling to traditional subsistence or industrial life-styles. The cost of housing, land, and services generally increases with an influx of immigrants from metropolitan areas. Local residents who cannot adapt to economic growth may suffer social and economic disadvantages. Restrictions on further industrial or residential growth can severely restrict economic opportunities for local residents who have relied on jobs in farming, forestry, fishing, or other extractive industries and must find new employment in construction and service industries. Political conflict between immigrants and 'locals' is particularly intense in rapidly developing nonmetropolitan counties throughout the West.

There are also indirect conflicts between residents of suburban and nonmetropolitan areas and the poor and ethnic populations in urban areas. Many people move to the suburbs or rural counties to live around people of similar income, education, and race. The social and economic homogeneity of suburban life has long been a concern of urban sociologists, and the 'white flight' that followed forced busing of urban school children simply reinforced this pattern. Lower income people cannot afford the price of housing, land, and transportation associated with suburban and exurban living. No longer the refuge for the poor, the forest on the residential fringe has become the refuge of the relatively affluent who seek the protection afforded by social homogeneity.

Consequences of Advanced Industrial Society for Population Redistribution

Unanticipated patterns of population redistribution have taken place in the last fifteen years (Kasarda 1980). Interregional migration has caused the South to gain population through net migration from the Northeast, and the West has gained through net migration exchanges with the North Central region. Within regions, nonmetropolitan areas ceased to lose population by the mid-1960s and have since exhibited rapid population growth through net migration from metropolitan areas. Within metropolitan areas, the population and job opportunities have shown a growing mismatch between places of work and places of residence: professional and white-collar jobs are located in the city center, while people with these skills live in the suburbs; blue-collar jobs are increasingly located in the suburbs, although people with these skills are drawn to cheaper housing and public transportation in the city center. The pattern of reverse commuting resulting from this mismatch creates the need for more transportation facilities. These redistribution patterns are associated with social and economic changes brought about by the emergence of advanced industrial society.

Substantial shifts in net migration from one region of the country to another reveal substantial changes in American society and its economic structure. Geographic settlement patterns and ecological linkages will be affected by these interregional shifts, since a large portion of residential growth takes place in areas covered by forest or other wild-land vegetation. Yet the reasons migrants are attracted to a region are as important as their contribution to population growth. Kasarda (1980, pp. 376-377) attributes Sunbelt growth to several economic and noneconomic reasons, most of which can be attributed to emerging patterns of advanced industrial society:

> (1) a growing footloose retirement population who have private pensions, Social Security payments and other sources of income not tied to specific locations and who have sought the milder winter climates that the South, in particular, provides; (2) the introduction and spread of central air conditioning systems that allow far more comfortable summertime living and working conditions; (3) life-style changes oriented to more recreation and year-around outdoor activities; (4) changing racial attitudes permitting blacks new opportunities to participate in the mainstream economic and social institutions; (5) a more progressive political orientation and an increasingly cosmopolitan atmosphere in the metropolitan South that has altered the antiquated 'red-neck' image many northerners and westerners had of the South; (6) generally lower land, living and amenity costs; (7) a dramatic improvement in the quantity and quality of consumer services brought about largely by rising personal income levels. . . .
>
> The South's attractiveness as an industrial growth pole is a function, in large part, of dramatically improved accessibility to national markets via new interstate highway systems, shifting energy sources and costs, and lower manufacturing wage rates interacting with a changing structure of the national economy and growing negative externalities of the concentrated northern industrial metropolis (e.g., congestion, powerful unions, high land costs, high taxes). To these technological and materialistic conditions are added healthy doses of progrowth attitudes and industrial boosterism on the part of the southern communities and their government officials.

Most interesting for our purposes is Kasarda's observation that 'growth in tertiary and quaternary industries (e.g., recreation, tourism, finance, research and development)' led manufacturing growth in 'the southern economy between 1970 and 1980, far outpacing any other region of the nation in job growth in these sectors' (p. 378). In other words, the characteristics of advanced industrial society have been important in drawing migrants and business to the South.

Nonmetropolitan Growth

The emergence of an advanced industrial society has also been accompanied by a sharp reversal in the long-established trend for nonmetropolitan regions to lose population to metropolitan areas. For almost a century, industrialization of the United States was accompanied by concentrated metropolitan growth and population decline in many rural locations. By the early 1970s, demographers had noted that migration from rural communities to cities had been superseded in magnitude by the migration of city dwellers to the urban fringe, rural communities, and isolated wild land settings (Beale 1975). Most significant is that nonmetropolitan counties gained population in all census regions, and that this growth has taken place in counties that are not adjacent to metropolitan areas, in addition to those adjacent. Between 1970 and 1980, nonmetropolitan areas grew by 32 percent in the West, 17 percent in the South, 13 percent in the Northeast, and 7 percent in the North Central region (USDA 1982). Net migration to nonmetropolitan areas is greatest in small towns and rural areas.

Kasarda (p. 381) identified some of the reasons for the unanticipated turnaround in net migration to nonmetropolitan counties:

> (1) the extension of the interstate highway system through many remote areas and the expanded all-weather surfacing of rural roads; (2) the expansion and linkage of locally owned telephone companies into the Bell system; (3) the extension of power lines and other public utilities and services, such as new, centralized, rural water systems throughout most nonmetropolitan counties; (4) telecommunication advances, including cable TV; (5) the proliferation of nationally subsidized standardized consumer goods and service establishments in nonmetropolitan counties; (6) dual residences associated with rising real incomes and increased leisure time; (7) a growing footloose retirement population; and (8) life-style changes oriented to bucolic, less densely settled environments.

Moreover, technological advances in communications and transportation enabled industries to loosen their ties to metropolitan centers and locate in remote areas where land and labor are cheaper and there is a lack of congestion. Although there has been substantial growth in manufacturing jobs in nonmetropolitan counties, growth of employment in the service sector has been greater—increasing sharply to almost two-thirds of all nonmetropolitan workers in 1980 (Kasarda 1980). People employed in agriculture make up less than 10 percent of the nonmetropolitan labor force. Hence, a large share of all nonmetropolitan employment is not tied to a particular location. Such territorially unrestricted sources of work and income further loosen people from traditional rural patterns that coupled nonmetropolitan living with the extraction of materials from land.

Intrametropolitan Redistribution

A far more familiar pattern of population redistribution is the spread of industrial and residential development to the suburban ring around metropolitan centers. At first such development was primarily residential use by professional and white-collar workers. However, advanced communications and transportation facilities make it more profitable for manufacturers, wholesalers, and retailers to locate in modern industrial parks and shopping centers along major suburban and exurban expressways. Advances in computer-aided communication technologies remove the need for administrative offices and research and development firms to be located in close proximity in concentrated developments. Decentralized offices and manufacturing facilities offer the attraction of environmental amenities, lack of congestion, and opportunities for many employees to reside near their place of work. In addition, many employers secure a better qualified and more stable work force by offering the amenities desired for a modern leisure-oriented life-style: campuslike settings with recreational facilities and restaurants provide relatively self-contained living environments.

The trend is for central city areas to house many of the blue-collar workers and those seeking unconventional life-styles. Industries still suited to a central location are banking and finance institutions, as well as convention services and the hotel, restaurant, and entertainment and cultural institutions that serve tourists. In summary, the suburbs and remote locations will continue to experience far more professional and white-collar job growth and residential growth than the central cities.

Consequences of Advanced Industrial Society for Ecological Linkages

Migration of people to the urban fringe and nonmetropolitan counties is accompanied by proliferation and growth of their linkages to wild land ecosystems. Concern with the relation between natural ecosystems and people has expanded well beyond the traditional concerns of forest influences. Water, climate, and fire are only a few of the many ways in which local residents are linked to natural systems in their immediate vicinity. In addition to their influences on microclimate and water quality, wild land ecosystems benefit people by producing oxygen, absorbing carbon and other wastes, buffering noises, providing food, fuel, ornamental vegetation, and scenery, and affording places for people to play and relax. Residents alter the structure and function of natural ecosystems by removing wood and other vegetation, polluting soil and atmospheric reservoirs with human wastes and household biocides, permitting household pets to act as predators on wildlife, redirecting the flow of water, lowering the water table, introducing exotic biota, and igniting unwanted fires. Wild-land ecosystems also constitute natural hazards for people in the form of dangerous or predatory animals, falling trees, and landslides, besides long-recognized hazards such as wildfires and floods (Meier and Ewald 1979).

We know relatively little about the nature and extent of these linkages. To illustrate our ignorance, I will briefly discuss interface wildlife hazards—a research problem to which I have devoted considerable time in the last three years. Contemporary settlement patterns, life-styles, and fire management practices could not be more effective in coupling wild-land fuels with fuels in residential structures. Migrants to wild-land areas prefer to build in the midst of natural vegetation in order to capture a sense of privacy. Architectural tastes dictate a 'natural' structure that fits into the landscape; cedar shake roofs, wood siding, wood decks protruding from the structure into surrounding vegetation, and large 'picture' windows facing natural views are all effective in linking structural fuels with surrounding vegetation. In addition, residential structures are located on small, winding, unmarked roads, and are supplied with only enough water to meet routine domestic demands.

To protect these residences, fire is aggressively suppressed in surrounding forest and brushland areas. As a result of successful fire suppression, the vegetative biomass accumulates to form a fuel complex far more explosive than would have been found under natural conditions where periodic fires would reduce accumulations of flammable vegetation. Uncontrolled wild-land fires tend to be larger and burn more intensely than they would with less fuel accumulation. As a result, structural losses and threats to life are greater.

Moreover, existing institutional arrangements do little to discourage people from building in hazardous environments or decoupling their structures from wild-land fuels. Mortgage money and fire insurance are available to people who build or buy in fire hazardous environments or require measures to decouple interface fuel complexes. When there are major residential losses to wildfire, the government provides emergency aid in the form of low interest loans to residents. Other than fear, there is little incentive for residents to change their choice of location, architectural taste, or domestic habits. We would have to exercise unusual creativity in order to do a better job of designing conditions ripe for disaster.

We are just beginning to assemble systematic knowledge of the interface fire problem. Far less is known about other important ecological linkages between residences and their wild-land environment. The harvesting and use of firewood by suburban and nonmetropolitan residents is another problem deserving immediate attention. Tree harvesting has ecological impacts on accessible forests, and the smoke from firewood consumption causes air pollution. Moreover, fire protection problems are created by improperly installed wood stoves. There is also a need for detailed knowledge of ecological linkages involving household pets, use of biocides, and hazards resulting from unmanaged residential forests.

IMPLICATIONS FOR FOREST POLICY

The conjunction of changing social values, population redistribution, and ecological linkages creates a complex environment for the practice of forestry. Foresters will need to rethink both the purposes of forest management and the means for accomplishing them in order to meet these challenges. I will conclude by discussing several implications for forest policy making of settlement patterns, ecological linkages, and political conflict on the residential fringe.

First, contrary to popular belief, conversion of land from forest land use to residential or industrial use does not necessarily contribute to shortages in wood supplies. Intensive management of high-site lands in regions not undergoing residential or industrial development can compensate for the loss of wood supplies in developing regions. The application of advanced technology in the form of fertilizers, biocides, and specially designed harvesting and regeneration systems can produce more wood from a smaller land area.

Second, economic growth of a particular region, and the nation as a whole, is likely to be enhanced by conversion of forests to more productive uses. The historical review has shown that forests have always grown on *residual land*—land that was not very valuable for other purposes. Population growth and redistribution have throughout history contributed to the clearing of forests and their replacement by other land uses. In the United States, relatively unrestricted land markets have resulted in widespread conversion of forests to more productive land uses. Some of the most productive agricultural land in the East, South, and Midwest was forested when settlers first arrived. Factories, highways, airports, and houses now occupy lands on which forests once grew. Recent patterns of population redistribution indicate that the land market continues to work well for facilitating locational decisions by industries and residents. Unlike many European nations, the United States has not restricted migration, with the result that people move from one place to another to avoid costly or unpleasant environments.

Third, forest lands, along with other wild lands, provide numerous ecological benefits at relatively low costs (Odum et al. 1976). Natural vegetation in a residential or industrial environment cleans the air, buffers noise, absorbs waste, provides oxygen, modifies the microclimate, offers natural beauty, and affords residents a recreational setting. Many of these benefits would have to be provided by expensive technologies to maintain an equivalent level of environmental quality. Thus, there is a need to consider the multiple ecological benefits and uses associated with wild lands. Foresters too often dismiss the contributions that natural vegetation can make to the quality of a residential environment.

Fourth, natural environments also present hazards for residential living. There is a need to account for the costs associated with protecting people from wildlife, animals, flooding, and landslides, since these costs are often shared with residents who do not occupy hazardous areas. Institutional arrangements need to be developed to distribute equitably the costs of protecting residents from natural hazards on the residential fringe. Hazard zoning, adjusted insurance rates, mortgage restrictions, and tax incentives are examples of institutional mechanisms to ensure that the people who choose to live on the residential fringe pay their full share of the costs associated with living in a more hazardous environment.

Fifth, the migration of people to exurban locations may be reversing long-established trends toward separation of the place of work from the place of residence. People who settle in new industrial developments or small towns often have the opportunity to work close to their residence. The geographic intersection or union of work and residence can have political significance for foresters and planners, since people may find it easier to mobilize political support for protecting the values they assign to the local natural environment. Just as the close proximity of work and residence aided the development of industrial labor unions, so the proximity of work and home may help people to build the social ties that will enable them to influence land use. Control over further development influencing zoning ordinances is only one of the many ways a highly mobilized local population can achieve what it wants. Limitations on the use of silvicultural practices and harvesting techniques can be effectively imposed by local residents who are united by a common concern with local environmental quality.

Finally, the problems of managing forests and wild lands on the urban fringe require specialized knowledge and skill that do not currently exist. The manipulation of natural ecosystems to produce a multitude of benefits requires not only scientific knowledge but also the skill to resolve conflicts between competing uses and to integrate a variety of management techniques to achieve special purposes. Foresters are perhaps the most suitable professionals for these tasks. Their general education and training in specialized techniques have enabled them to address complex problems in wild-land management. These same capabilities also suit them for solving problems of converting forests from wood production to residential environments and for continued residential use. The greatest challenge to foresters who seek to solve problems on the urban fringe will be to learn how to become effective agents for local residents, planners, developers, and environmentalists. This challenge will force foresters to rethink the purposes for which lands are managed, and to reintegrate those purposes with emerging forms of technology and socioeconomic organization.

REFERENCES

Beale, C. L. 1975. The revival of population growth in nonmetropolitan America. Economic Research Service Publication 605. U.S. Department of Agriculture, Washington, D.C.

Bradshaw, T. K., and E. J. Blakely. 1979. Rural communities in advanced industrial society: Development and developers. Praeger, New York.

Fernow, B. E. 1913. History of Forestry. University Press, Toronto, and American Forestry Association, Washington, D.C.

Kasarda, J. D. 1980. The implications of contemporary redistribution trends for national urban policy. Social Science Quarterly 61(3-4):373-400.

Lee, R. G. 1982. Alternative institutional arrangements for reducing fire hazards on the residential wildland interface. Final research report submitted to U.S. Forest Service, Pacific Southwest For. and Range Exp. Sta., Riverside, California.

Meier, R. L., and W. Ewald. 1979. Human habitat at the fringe of the forest: Character of the place. In: Our National Landscape, Proceedings of a Conference on Applied Techniques for Analysis and Management of the Visual Resource, pp. 738-743. Pacific Southwest For. and Range Exp. Sta., Berkeley, California. U.S. Forest Service PSW-35.

Nielson, J. C., B. Shelby, and J. E. Haas. 1977. Sociological carrying capacity and the last settler syndrome. Pacific Sociological Review 20(October):568-581.

Odum, H. et al. 1976. Net energy analysis of alternatives for the United States. In: Middle- and long-term energy policies and alternatives, Vol. 1, Appendix. Hearings before the Subcommittee on Energy and Power of the Committee on Interstate and Foreign Commerce, U.S. House of Representatives. U.S. Government Printing Office, Washington, D.C.

USDA. 1982. Handbook of Agricultural Charts. U.S. Department of Agriculture, Supplement to Agricultural Handbook No. 592. U.S. Government Printing Office, Washington, D.C.

Young, C. R. 1979. The royal forests of medieval England. University of Pennsylvania Press, Philadelphia.

Vaux, H. J. 1982. Forestry's hotseat: The urban/forest interface. American Forests 88(5):37, 44-46.

17

HUMAN NEEDS FOR RENEWABLE RESOURCES AND SUPPORTIVE ENVIRONMENTS

Rachel Kaplan

Many actions have unintended consequences. As we marched steadily across the land, pushing back the frontier and clearing for habitation, progress was visible. With land plentiful, trees abundant, and people relatively scarce, there was little reason to weigh silvicultural alternatives, to entertain master plans. But gradually, thanks to hindsight, development and technological innovations have come to be seen as complex. Many of the social problems we now face have been the consequences of what had appeared to be a virtuous drive toward progress. Unintended consequences often stem from a failure to anticipate the way people—individually and in groups— respond to change. Were we able to foresee what has so often been visible only retrospectively, perhaps the urban/forest interface would not be a catch phrase for a bundle of problems, but rather the label for an exciting series of opportunities.

The intent here is certainly not to decry the wonders of science or technology, nor to pretend that 'good' and 'old' have a meaningful historic relationship. What I would like to discuss, however, is the complex interface between the environment (i.e., forest, rural, urban) and the humans who sometimes fail to recognize the many ways in which they depend on it.

Humans have been available as a subject of study for quite a while. In fact, each of us has devoted much of our own life to understanding fellow humans (not to mention ourselves). Despite this extensive exposure, and considerable scientific investigation, we seem rather poor at predicting human behavior. We probably know more about intensive forest management than about circumstances that are supportive of human well-being. We have achieved remarkable progress in prolonging life; for many elderly people, however, physical health may be ahead of their sense of leading a meaningful existence or of making a difference in some larger way.

The relation between quality of life and the environment itself has received little analysis. From an economic standpoint, much human behavior may seem irrational. For example, the knowledge that one's community is to be flooded for reservoir development would suggest that one should accept the government's compensation and relocate. But as Johnson and Burdge (1975) have shown, long-time residents in such communities cannot tear themselves away. 'Uprooting' is a powerful image of the strength of the interface between environment and person.

Many other individuals—perhaps those who have had little opportunity for attachment to the land—are finding the move to new locations quite attractive. This trend has been of sufficient proportions that the conversion of forest land to residential and other land uses is the focus of this symposium. In the last decade, major shifts have occurred in choice of residential location. Although three-fourths of the population is still considered to be living in urban places, even within the metropolitan designation about one resident in seven is living in rural surroundings (Long and DeAre 1982). Moreover, in both metropolitan and nonmetropolitan contexts, growth has been inversely related to size of settlement—the smaller the population category (in census terms), the greater the rate of growth. This has led Long and DeAre to conclude that in the 1970s 'smallness came to be associated with growth' (p. 1112).

THE SOUGHT ENVIRONMENT

As many people rush headlong to new frontiers and settle the still unsettled areas, they come with hope and expectation. They are, no doubt, seeking many features that were lacking in the previous setting. The quest for less density—for fewer people and more land—is an expression of a variety of desires.

Analysis of people's interaction with the environment often focuses on activities and recreation. The outdoor environment is, of course, the context for many activities, and many of the most popular among them can be pursued in the forest context. They vary from active to less active to passive. They vary in consumptiveness, in the needs for equipment, and in whether they are engaged in alone or with others. While it is important to examine these patterns, the emphasis on activities, and on recreation, limits our understanding of the role the environment plays.

Much of the escape to the new frontier involves images of wide open spaces, trees and forests, and closeness to 'nature.' These are not images of activities, but of place. They involve perception of the environment itself. To consider these as 'passive activities' is to miss the point. While looking out the window at the 'endless' beyond is by some definitions passive, the environment itself, rather than the act of viewing, is what is important to the experience. This aspect of the forest is perhaps the most overlooked. It is difficult to define, hard to measure, and elusive.

To label this notion of 'environment-as-perceived' as 'aesthetic' or 'amenity' oriented suggests that we are dealing with aspects that are desirable, but hardly essential: in hard times, such luxuries as a nice view may be less vital. Such a perspective comes more readily to those for whom the proximity to natural elements can be taken for granted. The absence of contact with the natural world has a stronger impact than is generally recognized. That is not to say that the deep, dark forest is everyone's image of the Good Life. But the open, tamed woods have many advocates. Even though we may not be aware of the role trees play in our well-being, and find it difficult to verbalize the forest's psychological significance, the natural environment is nonetheless a vital aspect of effective functioning (Kaplan 1983).

While the environment-as-activity and the environment- as-perceived are necessarily interrelated, the distinction between them is nonetheless important. The focus on activity can blind us to the importance of the visual resource. Because of the emphasis on activity, the forest-in-the-wilderness and the forest-in-the-city are considered as largely dissimilar. Yet, one could readily take photographs of these two forests and find it difficult to distinguish them. One could spend time in the urban forest and have experiences quite similar to those afforded by the more pristine setting. Even the sound patterns may not be that different, for one is hardly immune to airplane noises in the wilderness and the trees may buffer the sound of the nearby expressway in the urban forest. It is even possible that one would meet no more people in the urban forest than in a remote-but-popular location.

While there are, of course, differences between these settings, perhaps the most important ones are psychological. The differences stem from one's *knowledge* of the extent of the two settings. Extent here is both spatial and temporal, including the knowledge both that the forest continues and that one's time in the setting is of a certain duration. This suggests that the urban/forest interface is not only an ecological, geographic, and political distinction (as Robert Lee has pointed out), but also a cognitive one. We are dealing with a state of mind as well as a state of reality.

The experiences that migrants to the new frontiers are seeking are closely related to this cognitive state. Just as the urban forest may be a microwilderness, the urban rural fringe is a little world which can seem far removed from the urban center that is only a few minutes away. Comparably, as people move to nonmetropolitan areas, they are seeking that *apparently* endless expanse of nonbuilt environment. The view is not, in fact, endless, nor does it remain unchanged. But the *perceived* density, the sense of closeness to land, and the feeling of coherence are important factors.

MEETING MULTIPLE NEEDS

Perhaps the new frontiers are meeting the needs of those who have migrated there. In the new setting, with a closeness to the land, with some of the advantages that smallness affords (Denman 1978, Gump and Barker 1978, Williams 1977), quality of life may be enhanced. But humans need more than a supportive environment in which to grow. To a much greater extent than most people understand, they are also dependent on the things that grow on the land. The need for food and fiber is not likely to decline as the population continues to increase.

It is reassuring to find analyses that proclaim that the loss of forest land is not a problem in terms of projected timber requirements. The technology seems to be available for producing more wood on less land, given the selection of the most suitable conditions. In fact, several chapters in this volume suggest that with intensive forest management procedures, land conversion is not a serious problem.

While it may be possible to produce these renewable resources using less land, it may not be desirable to do so. It is interesting to see cautions raised in recent issues of *American Forests* (Putz 1982) about the ramifications of new approaches to the production and harvesting of wood. While wood may be a renewable resource, forests may not be. The technology that leads to higher yields also draws on other resources—which are finite.

Forests are the source of more than timber. They are places, environments. While the wilderness is, fortunately, not everyone's idea of a desirable place, it is sought by more and more people. The wilderness has long been the theme of inspired writing by naturalists and poets. The encounter with the forest can serve a therapeutic need (Gibson 1979); it is a source of emotional satisfaction. Gradually, an empirical literature is developing to substantiate the psychological benefits of the wilderness experience (Kaplan and Talbot 1983).

Intensive forest management may well be a modern-day example of technological innovation that is far ahead of our knowledge of the interrelatedness of humans and the environment. Our ability to anticipate unintended consequences of such technology is deficient. Perhaps a touch of humility and a gentler attitude toward our life support system is appropriate.

Reverting to old ways is neither desirable nor possible. The points I am trying to make are not intended to advocate the status quo or the glories of times when population pressures were milder. We must pursue strategies that foster stewardship, approaches that recognize the fragility and limits of our existence. We must consider actions that preserve diversity of opportunities for unforeseen futures. Perhaps it is even possible that such approaches can nurture our psychological needs as well.

This would suggest that rather than permitting the reduction of land devoted to forestry, it behooves us to protect and maintain these holdings. The variety of strategies taken by the state of California to accomplish this (see Dean Cromwell's chapter) are farsighted and exciting. Not only have these approaches protected the forest for timber, but wildlife, fisheries, and water quality have also been improved (Pesonen 1982). Through zoning and a variety of incentives, as well as public acquisition (see Donald Miller's chapter), it is possible to reduce the conversion of forest to other land uses. Such strategies would seem of utmost importance for many states to consider in order to maintain the potential for sufficient forest resources.

The Forest in the Urban Setting

Another category of approaches to the interface problem might focus on 'bringing the resource to the people,' rather than having the people move to the resource. The phenomenon of ruining the very quality one is seeking is hardly a rare event. The wilderness that is available for all kinds of recreation cannot remain a wilderness (Hendee and Stankey 1973). The wide open natural areas which attracted the newcomers of the 1970s are quickly becoming a new wave of suburbs. What might be considered the ruralization of the countryside is a miniature version of urbanization. Or, as Gaines (1982) suggests, the move to smaller nonmetropolitan areas is perhaps more appropriately called 'deruralization.' The countryside is cleared to accommodate the very ones who sought it in its prior state. The woods one envisions as near enough to one's vacation home to provide ready access for hunting have become the backyard of the new neighbor's country home. Lee's 'last settler syndrome' reminds me of the definition of the conservationist as 'the guy who bought his cottage *last* summer' (Pieper 1974).

In order to 'bring the resource to the people' it is important to understand what the sought resource is. If people are fleeing the cities to live in the untamed woods, fleeing their neighbors to be isolated from fellow humans, fleeing from lawns and gardens to be surrounded by 'pure' nature, then bringing these qualities to the urban setting may indeed be difficult. But that is hardly the case.

The new setting is quickly tamed; lawns and gardens are landmarks of settlements, and the sought isolation is only relative, not absolute. To be near a forest, to have a view of things other than houses, and to have a place to grow some vegetables and flowers are desired features. To accommodate these desires need not require conversion of forest or agricultural land.

Part of the problem of the urban/forest interface, then, would seem to be related to a failure to appreciate the availability of the forest in the urban context. Lee's chapter reminds us of the strong ties forestry in this country has had to European forestry. The city forests in many European cities (Holscher 1973) provide sound examples of ways to bring the forests to the people—that is, to have them accessible while still managing them for multiple uses. Schabel's presentation (1982) concerning forests in Germany is particularly effective in identifying ways to accommodate a variety of activities—ranging from affiliative to solitary—within the urban forest.

American cities are not lacking in urban forests. While they are more often called arboretums or parks, in many cities there are sizable tracts of wooded land that are available to the public. I suspect we have not begun to appreciate the role these forests play. Their value may be recognized as an amenity reflected in the cost of adjacent homes. Their value is also emphasized by residents' outcries should even a portion of such forests be slated for conversion to other uses. But their role is far more pervasive.

Management of the natural environment in urban areas falls largely in the domain of 'parks and recreation.' This readily leads to two difficulties: consideration of the natural environment in terms of recreational activities, and a narrow view of the urban context. Examples of these limitations are provided by the widely used Recreation Opportunity Spectrum (Cordell and Hendee 1982). Six categories of settings are characterized in this spectrum. They range from 'primitive' to 'semiprimitive,' which is separated into motorized and nonmotorized, to 'roaded natural,' 'rural,' and finally 'urban.' Unlike the more 'natural' areas, the urban setting is characterized as at most having 'natural-appearing elements.' Such settings are oriented toward 'experiencing affiliation with individuals and groups' (as opposed to 'isolation from the sight and sound of others'). 'Experiencing natural environments, having challenges and risks afforded by the natural environment, and the use of outdoor skills are relatively unimportant.' If this description is veridical, it is no wonder that people who have the means would flee the city to find any 'real' nature. Are the trees in the city merely 'natural appearing elements'?

Such a characterization is detrimental in a variety of ways. Urban forests are not appropriately described as arenas for spectator sports or noisy places for loud gatherings, nor are they only witnessed as one speeds through in a motorized vehicle. While an urban resource must be managed in ways that can accommodate heavy use, the European example makes clear that this can be done without eliminating the natural resource. To consider the urban forest as not natural (or only 'natural appearing') because it has paths and benches is to fail to understand the human experience of the setting. True, the pristine wilderness is sought by some for its lack of visible management, but to many others the restorative capacity of encounters with nature is quite independent of such signs.

The other shortcoming of the way we approach the urban natural resource is that it is seen in recreational terms. Many of the ways the natural environment affects our well-being cannot be subsumed under this heading. The factory worker who rushes off during the lunch break to find a tree to lean against while consuming a sandwich is not 'recreating' in any usual sense of the word. But the tree is hardly a randomly selected setting. People who take longer routes to work (by car, or on foot or bike) to have a 'greener path' are also not partaking of recreational activities. Comparably, the view from the window is an important part of life. When the view is of natural elements—trees, fields, woods—it has a strong effect on neighborhood satisfaction (Kaplan 1983).

Thus much of the significance of the natural world in the urban context is as a perceived environment and as a mental state. The knowledge that the forest exists, that there is a park nearby, that the tree one planted will bear fruit, is sustaining and satisfying. But such flights of fancy are not limited to the urban setting. The newcomer to the rural fringe enjoys the knowledge that now there is an endless uninterrupted vista—even as the curtains are drawn and a new industrial complex is erupting in the distance. The wilderness too provides such cognitive satisfaction for the 'vicarious visitors' for whom knowledge of its existence is a source of strength (Hendee and Stankey 1973).

The cognitive satisfaction the forest provides is of major importance. It is difficult to fit into a rational analysis of human decisions. The often expressed frustration that the small woodlot owner is not economically oriented provides a persisting example. As Miller indicates, many of these people express little interest in the monetary gain they could reap from their land: many of their satisfactions come from the perception of their forest and from the knowledge that it is theirs. But these same forests, although not in the public sector and not contributing to timber harvest, are no doubt providing some satisfaction for others as well. The drive through the countryside, the walk on adjacent lands, and the knowledge that there are woods nearby all contribute to well-being.

To summarize, then, there are important ways in which the search for 'somewhere else' could be met 'right here.' If the nearby setting was recognized as a vital component of the supportive environment in the urban and suburban context, land would not be needlessly converted to other uses. At the same time, the well-being of the urban population could be enhanced where the population is already concentrated. The natural environment plays an important part in providing such supportive and restorative elements, and it does so in ways that are not recreational.

THE SEARCH FOR ANSWERS

This analysis of human needs and satisfactions relies heavily on information. It is the knowledge, the familiarity, the information we have about things in our world that is central to human functioning. The lack of such comprehension, the inability to grasp what goes on around one, is the source of acute frustration, suffering, and hostility. But the information one already has is easily taken for granted. While the desire to be able to predict or to know what to expect is pervasive, the yearning for exploration, for the unknown, is equally intense. Thus, in many circumstances, humans teeter between satisfaction with understanding and the desire to extend their knowledge, between the quest for certainty and the flirtation with uncertainty. New challenges are the lifeline of an information-hungry animal; the failure of comprehension may be its downfall (Kaplan and Kaplan 1978, 1983).

Information is central to our comprehension not only of the environment but also of the future. It is hardly surprising that two kinds of recommendations are offered for many problems. One involves the need for educating and the other the call for further knowledge, for research. Perhaps there are circumstances where ignorance is indeed bliss. But this bliss does not extend to the inequities that follow from uninformed decisions, or to coping with a people-rich and resource-declining world. Regretted consequences are unlikely to be avoided without a concern for improved knowledge.

Knowledge Dissemination

While there are certainly many areas that require further knowledge, there are many others where knowledge is at hand and the need is to deliver it. One such area involves the public's ignorance about the forest. Many citizens have no grasp of the relation between trees, wildlife, and water; they know little about sustained yield or about the extent of need for forest products. Another area where further education is needed involves the professionals' understanding of the importance of the natural environment to psychological health. Behavior that may seem irrational in economic terms may be quite reasonable with a different understanding of the human animal.

If such knowledge is at hand, its dissemination would seem to be a relatively forthright task. The fact that we so often fail in efforts to disseminate knowledge—to educate—is striking. Unfortunately, the delivery of information and its reception are by no means equivalent. While there is no question that efforts must be made to educate the public—and the professional too—it would be irresponsible to make such a recommendation without acknowledging the difficulty of the task.

This is not the place to deal with means of enhancing communication, with ways of making information more available to the recipient (Kaplan and Kaplan 1978, 1983). It is important, however, to recognize how easily we fail in these attempts and that the failures are themselves understandable. Much as humans crave information and new frontiers, they also readily ignore information that they think they already have or that does not fit their knowledge structures. At times it seems surprising that communication works at all!

Knowledge Acquisition

There is a considerable industry devoted to generating answers to questions. Research and development aimed at intensive management may be expensive, but it is relatively straightforward and easily justified. By contrast, research on the psychological and sociological aspects of land use is nebulous. 'People' research is fraught with difficulties; the lack of such research, however, is fraught with dire consequences.

We have devoted much effort to understanding the requirements for a supportive environment for zoo animals and for laboratory animals as well. The requirements extend well beyond the provision of food and standards for cleanliness. If we were to achieve comparable understanding of the properties of supportive human environments, I suspect the social/environmental problems of the contemporary world would be far fewer.

One of the areas about which we know relatively little and which would seem particularly pertinent to the urban/forest interface deals with the question of extent, both actual and perceived. There are interpretive trails which marvelously achieve a sense of a large meandering region although, in fact, the area is small. Japanese gardens also accomplish a sense of space far greater than the acreage would suggest. In a humorous vein, Frisbie (1969) offered suggestions for miniexcursions into the close-to-home natural world that generate equivalent 'red blood density' to the densifying powers of the 'real' wilderness. For purposes of experiencing the forest, how vast must it be?

The forest is part of the visual resource that several federal agencies are charged with managing. Invariably, diversity is included in the visual resource management guidelines as an important component of scenic quality. Diversity, with respect to the forest, presumably involves more than species diversity. The juxtaposition of woods and open areas, the mix of dense and more spacious forest, the diversity in the shape of the trees themselves, all play a role in environmental preference. But we know little about these.

Our knowledge is particularly lacking with respect to anticipating future regrets. Many proposed changes are perceived as threatening and some are perceived as improvements. The retrospective view of change, however, may be quite different. I suspect that many changes would never have been undertaken if there had been a better way to simulate the consequences. It is all too rare that the trade-offs are clearly weighed when changes are promoted. If people were asked whether they would like more jobs *and* the concomitant pressures that growth would bring to their community, the land conversions of the past few decades might have taken a far different course.

The Forest Service has recognized the need for an active research program focused on the urban forest (Moeller 1981). Fortunately, some of this effort is devoted to understanding the importance of the urban forest in social terms. While such research is at the mercy of federal spending priorities, it is noteworthy that its urgency has at least been acknowledged.

Much remains to be explored to enhance our understanding of the forest as a place, an experience, quite apart from the context of activities. The forest is a place, the garden is also a place, even a single tree at the curbside is experienced as a vital element of the natural environment. The comprehension of this relationship to the natural world is essential if the forest is to remain a resource for future generations.

REFERENCES

Cordell, H. K., and J. C. Hendee. 1982. Renewable resources recreation in the United States: Supply, demand and critical policy issues. American Forestry Association, Washington, D.C.

Denman, C. C. 1978. Small towns are the future of America. In: S. Kaplan and R. Kaplan, eds., Humanscape: Environments for people, pp. 274-279. Ulrich's, Ann Arbor, Michigan. Reprinted, 1982.

Frisbie, R. 1969. It's a wise woodsman who knows what's biting him. Doubleday, New York.

Gaines, T. A. 1982. Deruralization. Farmland (Newsletter of American Farmland Trust), September.

Gibson, P. M. 1979. Therapeutic aspects of wilderness programs: A comprehensive literature review. Therapeutic Recreation Journal 13(2):21-33.

Gump, P. V., and R. C. Barker. 1978. Big school, small school: Overview and prospects. In: S. Kaplan and R. Kaplan, eds., Humanscape: Environments for people, pp. 407-412. Ulrich's, Ann Arbor, Michigan. Reprinted 1982.

Hendee, J. C., and G. H. Stankey. 1973. Biocentricity in wilderness management. BioScience 23(9):535-538.

Holscher, C. E. 1973. City forests in Europe. Natural History 82(4):52-54.

Johnson, S., and R. J. Burdge. 1975. An analysis of community and individual reactions to forced migration due to reservoir construction. In: D. R. Field, J. C. Barron, and B. F. Long, eds. Water and community development, pp. 169-188. Ann Arbor Science, Ann Arbor, Michigan.

Kaplan, R. 1983. The role of nature in the urban context. In: I. Altman and J. F. Wohlwill, eds., Behavior and the natural environment. (Vol. 6 of Human Behavior and Environment). Plenum, New York.

Kaplan, S., and R. Kaplan, eds. 1978. Humanscape: Environments for people. Duxbury (Division of Wadsworth), Belmont, California. Ulrich's, Ann Arbor, Michigan. Reprinted 1982.

Kaplan, S., and R. Kaplan. 1983. Cognition and environment: Functioning in an uncertain world. Praeger, New York.

Kaplan, S., and J. F. Talbot. 1983. Psychological benefits of a wilderness experience. In: I. Altman and J. F. Wohlwill, eds., Behavior and the natural environment. (Vol. 6 of Human Behavior and Environment). Plenum, New York.

Long, L., and D. DeAre. 1982. Repopulating the countryside: A 1980 census trend. Science 217(September 17):1111-1116.

Moeller, G. H. 1981. Research programs in urban forestry. Trends 18(4):4-9.

Pesonen, D. E. 1982. California forest law shows value of environmental regulation. Ann Arbor News 9/28/82.

Pieper, L. P. 1974. Lake development: Problems and possibilities. Master's thesis, Landscape architecture, University of Michigan.

Putz, G. 1982. Dragons in the woodlot. American Forests 88. Three-part series, August, September, and October.

Schabel, H. G. 1982. Urban forestry: Some lessons from Germany. Presented at Society of American Foresters National Convention, Cincinnati, Ohio, September 21.

Williams, H. S. 1977. Smallness and the small town. Small Town 8:7-15.

18

THE NEW PEOPLE IN THE WOODS

Edward J. Blakely

The rapid alteration of the American population distribution from snowbelt to sunbelt and from urban to nonmetropolitan has become a topic of increasing interest. Underlying the analysis of this population change is the theme that individual life-style choice is a larger component in the shift than economic determinism. Robert Lee, like many other social scientists, has developed his own analytical framework and perspective on the issue. In his analysis he uses land use of wilderness areas as a central tenet in arguing that a new social order is emerging—one that is not bound by previous historical realities. The advanced industrial society, Professor Lee suggests, uncouples work from specific location. This allows many individuals, perhaps too many, to select their fantasy environment as residence and work place. As a consequence, fragile wild lands such as forest, brush, deserts, and marshes are threatened by man's extension of his domain. Professor Lee's arguments are well developed with respect to this proposition. Thus, I would like to offer a complementary discussion on some of the characteristics of the emerging socioeconomic patterns in an effort to explore potential public policy directions or alternatives.

The emergence of the advanced industrial society represents more than population shifts and the demise of the manufacturing industry. New social values and institutions are rising while the traditional order is under severe pressure to transform itself or cease to exist.

Between 1960 and 1970 most rural areas experienced population decline and economic depression or stagnation. Almost four million people left rural areas during that decade; at the start of the period, one-third of the nonmetropolitan population (33.2 percent) was below the poverty level compared with 15.3 percent in metropolitan areas. By 1980 the situation had changed. In the last decade three million persons migrated to nonmetropolitan areas, and the number of persons below the poverty level had declined to 15.4 percent (Haren and Holling 1979, Bradshaw and Blakely 1982). In addition, from 1962 to 1978 nonmetropolitan areas added 8.5 million nonagricultural jobs, a 65 percent increase during this period. Total personal income in nonmetropolitan areas increased 84 percent during 1970-76, while the increase in metropolitan areas was only 70 percent. This growth in income improved the nonmetropolitan economic base. The new migrants are a new source of economic growth because of their intellectual capacities. As shown in table 1, many newcomers to nonmetropolitan areas are very well educated. This growth in educated population, employment, and income in nonmetropolitan areas during the 1970s has dramatically altered many policy factors associated with the concepts Americans have of rural areas, wild lands, conservation, and historical preservation. These trends are moving so rapidly and dramatically that Professor Lee has been unable to sketch the dimensions and possible impacts in a single chapter. Among the trends that I wish to explore that have a bearing on the forest/wild land discussion are those related to changing location, new life-styles, new rural community configurations, and altered public policy systems.

Table 1. Educational Level of Newcomers.

Level	Working (%)	Retired (%)	Total (%)
Less than high school	4.7	28.1	12.9
High school graduate	16.8	29.7	21.2
College, some B.A.	63.0	34.4	54.0
M.A., Ph.D.	14.8	6.3	11.2

Source: Blakely and Bradshaw 1980.

CHANGING LOCATION

The American culture is deeply affected by its historical mystique with respect to the open country. This mythology has so captured the national psyche that television advertising and other media capitalize on the open country and small town themes almost as frequently as on sex and status. Whether because of this depiction of Americans or as a result of deeper historical trends, the majority of Americans report a desire to live and work in a small town. Obviously, in one of the most urbanized and affluent countries in the world, opportunities to live out this fantasy have to take many different forms. Consequently, we find millions of dollars and thousands of acres of land devoted to second homes in the country and hobby farms. Until very recently these opportunities for encounters with nature were sufficient. However, the advent of more transportable technology and new production systems has brought the factory right to the fields.

In fact, nonmetropolitan America is the only segment of the national economy continuing to grow. This growth, until recently, was phenomenal. These gains have not been in fields related to the land or its resources. They are jobs that provide a transfer of services to people. For example, services have increased employment by 6.2 million jobs and account for 66 percent of all nonmetropolitan employment. The vast majority of these jobs and industries went into the communities or areas foresters and other environmentalists consider most fragile. Most of the new firms were in open places of 'high scenic value' (near trees, lakes, deserts, or brushlands, etc.). It is precisely because these places offer the closest match between the ideal living arrangement and job opportunities, even at modest wages, that millions of people were attracted away from the cities.

The pressure of this form of economic transfer of plants away from the cities presents new dilemmas for everyone. While Professor Lee outlines the fire and similar dangers this situation poses, the more fundamental issue of the viability of such a juxtaposition remains to be confronted. This issue is of greater consequence in the advanced industrial context for two particular reasons. First, as the value of scenic places increases, factories will move into the woods. This is already occurring in the West Coast electronic industry in Northern California and Oregon. In addition, the potential for individual entrepreneurship to operate cottage industries in wooded areas has grown substantially. This industrialization of forest land (public or private) brings all the problems of urbanism into a landscape ill suited to cope with them. Some reports indicate that the computer software industry, which is expanding at 10 percent a year and will require about 1.5 million new employees between now and 1990, will accomplish most of this growth in highly scenic nonmetropolitan environments (Hall et al. 1982).

Second, the urbanization of small towns near forest areas brings new potentials and problems. Urbanites use natural resources and beauty but have little patience for preserving and managing these resources properly. That is, the urbanite wants to jog, maintain domestic animals, barbecue, and dispose of waste based on urban systems. Clearly, as these areas become more abused, their value deteriorates. But the footloose pattern of urban industrialism can traumatize forest and other areas beyond the point of repair.

LIFE-STYLE

The advanced industrial society presents new options for individuals to select their own life-style opportunities. Life-style in this context means everything from dwelling and work places to selection of entertainment and associations. The opportunities presented are so myriad that an individual can be part of various cultural patterns simultaneously; for example, vegetarianism can be combined with ardent game hunting. Most individuals and groups claim the new liberation of choice is welcome. But freedom of choice for an individual may limit the rights and opportunities of others, particularly animal life. The new freedom of advanced industrialism is the ability to form transportable urban cocoons (Bradshaw and Blakely 1978). As a consequence, many millions of urbanites with mobile vans can experience rural areas, forest, and wild land without missing the comforts of urbanism. Similarly, the migration to rural areas is supported by rapid improvements in electronic technology that allow quick access to urban health care, entertainment, and jobs. While all this is laudable in most respects, it has one overwhelming flaw: there is no restriction, beyond price, to the possession of this technology. Thus, individuals with little respect for or knowledge of wild lands have access to them without the prerequisites of education, training, or experience. As Dean MacCannell points out in *The Tourist* (1976, pp. 80-81):

> Modernity is transforming nature from a cruel alternative to community life into a place of play.... Powerful human passions evoked by nature were once available in a wide variety of situations: in the hunt, in the forest on the edge of camp, at sea beyond the horizon. The human group could, and did, draw heavily on the unknown forest and sea for inspiration in the creation of social solidarity out of opposition of man and nature.... Modern tourism is reorganizing nature and the touristic experience of it so it may continue to serve as a basis for unity in the family of man. The modern touristic version of nature treats it not as a force opposing man, something

Forest and wild lands practices must be sensitive to these groupings, feelings, and perceptions regarding the landscape. In some instances, these forest areas are viewed as sacred and in others act as cover for marijuana crops. Forest and wild-land management in these circumstances must be cognizant of the people as well as the trees.

Each community type presents different challenges. The blanket development of policies or regulations to preserve natural resources without an understanding of these issues will cause greater problems than those resolved. Therefore, new policy instruments are required—instruments that are designed for the physical settings of rural America and not urban systems modified for nonmetropolitan areas.

ALTERED PUBLIC POLICY SYSTEMS

Slowly and almost imperceptibly the public policy formulation system is evolving toward the advanced industrial society model. Party politics is giving way to interest group concerns, and ballot measures are replacing legislative deliberation. People are voting increasingly with their feet or their purses. The advent of mass communication makes it almost impossible for leadership to emerge and provide definitive guidance in any area. The issues are too complex and the system too broad for simple solutions to work effectively. Consequently, more groups seek solutions to issues through their favorite legislative device. This ranges in the wild lands arena from protection of river systems from dams to antilitter laws requiring returnable bottles and cans. It can be argued that the environmental protection argument would never reach public debate or legislative action without popular movements. The other side of the coin is that legislative and executive leadership waits increasingly for popular sentiment before taking required actions, particularly unpopular legislation. For example, the requirement in building codes not to use shakes or wood roofing materials in forest areas failed to pass in Los Angeles County when emergency legislation lapsed.

It is unfortunate that information on the needs of the nation's natural habitat is not disseminated widely. Public debate is too often carried on in black and white terms rather than in long-range resource development strategies. However, until a coherent policy development system emerges for natural resource management, the policy process will remain both diffused and confused.

Public Policy Directions

In order to develop policy alternatives or directions, some assumptions must be made. The basic assumptions arising from Lee's paper are:

1. The pressures on wild lands will grow rather than diminish, because of population growth in rural areas as well as recreational demands.
2. The desire of the American public to preserve forest and natural habitats cannot be fulfilled unless new management tools are developed.
3. The problem is not a problem of insufficient wood production to meet national demands; rather the issues related to land management reflect cultural and social needs.
4. Urban life-styles and housing patterns in wild lands represent a substantial danger to these lands.

Designing policy from these assumptions is particularly difficult because they represent conflicting concepts and policy issues. There is no easy formula to resolving these problems. Yet some system or process needs to be adopted that provides for orderly adjustments among conflicting requirements. Several proposals based on this premise are outlined below.

Wild Land Designation and Separation. There are already fairly well-regulated forest areas designed to manage certain forests or other areas. It would seem appropriate, however, to make graded designations of forests ranging from primitive preserves to hobby farm zones. In the most delicate areas only licensed individuals could enter for recreation or other purposes after specific training in wild-land ecology. In areas that are considered more accessible, general zones should be developed indicating construction type and activities that can be conducted. Wide bands or buffer areas should be developed between accessible and inaccessible areas. These buffer zones should be managed wildlife areas specially maintained to meet animal needs and to act as recreation sites.

Community and Wild Lands Protection. Every year thousands of acres of forest land are destroyed by fire. In addition to the destruction of forests and wild animals, homes are destroyed. In part, the location of the homes in or on the edge of forest lands contributes to the fires. These conditions could be significantly reduced if stronger measures in home construction regulating both location and building materials were taken. Cluster development and settlement should be required rather than scatter site development. States should be encouraged to develop land permitting processes rather than urban zoning to facilitate appropriate development strategies.

User Pays. Professor Lee suggests that the costs of fire hazards, floods, and similar locationally induced problems are borne by all taxpayers. Hazard zoning which incurs heavier tax rates should be imposed on residential or commercial properties. In addition, insurance rates and other costs should be higher for individuals living in hazardous or environmentally sensitive areas.

Urban Regional Forest. One of the reasons for the substantial demand for both private and public forest lands is the absence of suitable wild lands near the cities. Rather than attempt to accommodate more individuals in wild-land areas at considerable distance from urban centers, regional park systems with camping facilities should be developed close to urban areas. These facilities should be developed complete with second-home living areas. The areas could be developed specifically for recreational purposes complete with developed woodlands or forest areas.

As the population increases in nonmetropolitan areas, wild-land recreational sites should be developed on a regional basis designed to concentrate population in managed forest areas.

Foresters' Roles

Professor Lee suggests that foresters can and should assist communities to develop mechanisms to deal with policy issues. While forestry training may well have altered to include more study of policy issues, there is no evidence that group skills and communication are subjects strongly represented in the curriculum. Working with people, particularly in conditions of stress and confusion, requires specific skills. Furthermore, there are other developers and policy scientists with the requisite training to assist in this area. There is legitimate doubt as to the ability of foresters to provide social conflict resolution for society, inasmuch as their training continues to be in natural rather than political and social sciences.

REFERENCES

Blakely, E. J., and T. K. Bradshaw. 1980. The impact of recent migrants on economic development in small towns. Southern Quarterly 19(1).

Bradshaw, T. K., and E. J. Blakely. 1978. Policy implications of California's changing life styles. Institute of Governmental Studies, University of California, Berkeley.

———. 1981. Resources of recent migrants to rural areas for economic development: Policy implications. Division of Agricultural Sciences, University of California. September.

———. 1982. The changing nature of rural America. In: W. P. Browne and D. Hadwiger, eds., Rural policy problems: Changing dimensions. Lexington Books, Lexington, Massachusetts.

Hall, P., A. Markusen, R. Osborn, and B. Wachsman. 1982. The California software industry: Problems and prospects. Institute of Urban and Regional Development, University of California, Berkeley.

Haren, C. C., and R. W. Holling. 1979. Industrial development in nonmetropolitan America. In: R. Lonsdale and H. L. Seyler, eds., Nonmetropolitan industrialization, pp. 13-45. Halstead Press, New York.

MacCannell, Dean. 1976. The tourist: A new theory of the leisure class. Schocken Books, New York.

PART VI

COURSES OF ACTION

19

STRATEGIES FOR DEALING WITH THE URBAN/FOREST INTERFACE: THE RECENT CALIFORNIA EXPERIENCE

Dean A. Cromwell

In the last decade, forest management and use have been the subjects of heated political controversy. Nationally, this was reflected in the National Forest Management Act and various proposals to resolve the wilderness question. Equally bitter debate occurred in California over the Z'berg-Nejedly Forest Practice Act of 1973 and again in 1975 when thirteen hundred loggers in woods regalia and logging trucks descended on the state capitol to protest added state requirements.

The bitterness of this debate reflects a basic disagreement among segments of society about how forests should be used. These differences in values characterize all political levels—local to national—and define the 'urban/forest interface.'

The term 'urban/forest interface' describes a political phenomenon of conflicting values. Vaux, for example, has observed that the term refers specifically to a confrontation of political forces, all having different views about the value of forests to society (Vaux 1982). Some persons may see intensive logging as essential; others may believe that logging is inappropriate because of amenity, recreational, or watershed values.

These conflicting views often reflect disagreements about income distribution between segments of society and between our own and future generations. Forests represent capital to be converted to dollars or a resource to be enjoyed for noncommodity values. In either case, how society chooses to use its forests results in a distribution of the income to be derived from forest use.

In addition, forests represent investment capital for future generations. How forests are managed is thus partly determined by distribution of the future benefits of the forest.

Arguments over income distribution are by their very nature political. Questions of efficient allocation by a market mechanism cannot be settled until an income distribution is chosen. Consequently, the challenge of the interface is largely a political one of how to resolve conflicting values and associated income distributions.

Vaux also makes the point that the term 'urban/forest interface' does not represent a geographic phenomenon because the problems are not confined to a particular place. This is an important distinction. Usually the term calls to mind some specific geographic location where forest management and urban development occur together. It may even conjure up memory of a particularly heated squabble between loggers and nearby homeowners who did not want to see some trees cut. The problem is that visualizing the urban/forest interface in a geographic context can narrow thinking about strategies to deal with the interface. Basic strategies must address people's value perceptions, must find ways to facilitate communication between conflicting views, and must provide ways to resolve differences.

At the same time there are certain urban pressures on forest land areas that must be confronted just to preserve the forest resource. These effects include higher land values, fragmentation of ownership, and land use problems associated with more people living in forested areas. Unchecked, they operate against continued maintenance of a forest base. On the assumption that society wishes to preserve substantial amounts of forest land in good environmental condition, either for producing timber or for providing backyard scenic amenities, strategies must also be developed to deal with these pressures.

This chapter addresses both the urban/forest interface and urban pressures related to it. The urban/forest interface is treated as a political phenomenon, and general strategies are presented in such a context. Certain urban pressures are also described and approaches are outlined to deal with the effects. Emphasis is placed on the role of state and local government. Although actions of governmental agencies on federally owned forest lands are sometimes catalysts for political disagreements over the use of forests, strategies that are utilized or can be utilized by these agencies are not discussed.

Approaches may differ substantially between states and localities. They depend primarily on the values of urban populations and the political structure and traditions through which the values are expressed. However, for purposes of discussion, examples are drawn from California. California is a good example of a highly urban state that is also blessed with a rich forest resource. Ninety-four percent of its citizens live in cities or suburbs, but rural areas are growing faster than urban areas. Attitudes and perceptions are distinctly urban. Both the urban/forest interface and critical impacts of urban pressure on the forest resource are very evident.

In addition, in the last ten years California has adopted a number of new programs related to the urban/forest interface. These programs present a relatively coherent approach to the problems of the interface. The efforts, totally or in part, can be viewed as an innovative experiment that merits discussion as part of strategies to deal with the interface.

THE INTERFACE AS A POLITICAL PHENOMENON

Forests can have many uses. Not all of them are compatible. Given a fixed forest resource base and more people making demands on that base, intense conflicts can arise over how forest land is to be used. Historically, such differences in view have long existed. What is new, as Vaux points out, are the intensity of the controversy and 'the fact that the issues tend to become ever more sharply focused on particular pieces of forest land' (p. 37).

Strategies for dealing with the interface consequently must focus on ways to resolve political conflicts over differing value perceptions. They must recognize that values are developed in an urban environment and are expressed as such in the political process.

Approximately 75 to 80 percent of Americans live in cities or a suburban environment. People are removed from direct contact with the forest resource. Opinions are gathered from newspapers, television, and other media. The result, according to Henry DeBruin (1982, p. 19), is that 'few people experience the land and its resources first-hand, and fewer understand the critical role the land and its resources play in undergirding our lives.'

This is especially true in California. Most Californians live in cities and suburban areas. They tend to have only a limited understanding of forestry, largely derived from the news media. In general, Californians tend to be concerned about environmental protection. Vocal segments are sensitive to key environmental and political issues, such as increased offshore oil drilling disposition of public lands.

The structure of the political process reflects this predominantly urban nature. Social values are expressed politically through interest groups. Issues tend to become intense before they are noticed and dealt with by the political process. Increased intensity, particularly coverage by the news media, is essential to capture attention in a political process caught up in many controversial issues. For this reason, issues that focus on a particular piece of land actually can carry substantial symbolism and political significance.

Resultant political pressure often takes the form of working against the traditional land use, such as timber growing. It seeks to regulate industry by putting industry on the defensive and placing government in a mediating role. This is a task for government different from its traditional functions of aiding research and education and fostering cooperation.

Given the urban nature of the population and the characteristics of the political structure, strategies dealing with the interface should actively address urban perceptions about forest resources, increase foresters' knowledge of urban attitudes, establish a policy-planning mechanism to anticipate and focus on potentially controversial issues as soon as possible, and develop a methodology to identify and address site-specific impacts of land uses before they occur.

Actively Addressing Urban Perceptions

In the long run, various forest policies will only be as strong as the public's understanding and support. The purpose of actively addressing urban perceptions is to improve the public's knowledge of forestry problems and the importance of dealing with them.

In the last four years, California has taken several steps that directly or indirectly address urban perceptions of forestry and other renewable resources. Initial efforts concentrated on improving productivity of the state's private forest land and establishing an urban forestry program. Later efforts have been directed at all of the state's renewable natural resources. Under leadership of the Resources Agency, they represent deliberate attempts to reorder priorities of state government to give more attention to investment in renewable natural resources.

As such, these efforts address implicit urban perceptions that forests and other renewable resources will somehow take care of themselves and will always be available. A different message is put forth: the public cannot take renewable resources for granted; wise investment in and management of these resources, including forests, is the key for survival. This message has been widely circulated in the media and the legislature. To date, it has been well received, as programs and budgets have been passed in the legislature by large bipartisan votes.

These efforts also reflect a strategy to focus public attention on the need for sustained funding and long-term planning. Examples of sustained funding include the Forest Resources Development Fund, the Renewable Resources Investment Fund, the Geothermal Resources Fund, and the Energy and Renewable Resources Fund. The Forest Resources Development Fund provides funds for reforestation on private land; dollars come from sales of timber on state forests, reflecting the principle that income from the sale of the timber should be reinvested to improve forest productivity. The Renewable Resources Investment Fund provided $10 million to develop wood energy and other renewable resource programs. The Geothermal Resources Fund commits 30 percent of the income from federal geothermal leases to be reinvested in renewable resources. The Energy and Renewable Resources Fund assigns a portion of tidelands oil revenue income, up to $120 million a year, to reinvestment in natural resources; the premise is that returns from nonrenewable resources should be used to enhance renewable resources.

Long-range planning is dealt with in a twenty-year plan published by the Resources Agency. Entitled Investing for Prosperity, the plan sets forth long-range goals for renewable resources and programs to achieve these goals (California Resources Agency 1981). It 'presents a strategy for resources security in the next century and beyond through wise use of our remaining nonrenewable resources and the restoration and maintenance of our renewable resources.' Aggressive investment programs are set out for forestry, fish and wildlife, water, soils, parks and recreation, and coastal resources.

The state's urban forestry program, authorized in 1978, provides grants for management, planting, and maintenance of trees in urban areas. Urban residents, community groups, and local businesses cooperate to improve and beautify their neighborhoods. In the first three years of the program, the Department of Forestry estimates that almost 100,000 people planted almost 100,000 trees to enhance the quality of their communities (California Department of Forestry 1982a). Efforts have also been made to educate the public about the impacts of urbanization in forested areas. The program is creating an urban political clientele more knowledgeable about forestry and conservation. There is a growing dialogue between Department of Forestry personnel and community groups in urban areas. Ultimately, this is intended to provide a forum to discuss differing concepts of forestry and how forests are to be used.

In the context of strategies for the interface, these efforts are examples of attempts to raise the knowledge of the public about forests and other renewable resources. It is also important to continue building the public's information and understanding. One way to do this is to build a mechanism that provides credible information in a timely manner. California has no such device. This has led the Board of Forestry to propose creation of a Council on Public Education in Forestry (California State Board of Forestry 1981). The council would consist of representatives of each major group with concerns over the use of forests; it would develop, coordinate, and support an adequate program of information dissemination to the public. If successful, it could maintain the public's interest in forestry-related matters and could provide the basis for better discussion of differing views about how forests are to be used.

Increasing Foresters' Knowledge of Urban Attitudes

The profession of forestry has evolved in Europe around the idea of providing for wise management of forest resources. Today in some European countries, foresters are essentially responsible for most forest management decisions and are trusted by the public to make appropriate choices between timber harvesting and nontimber values.

This tradition suggests that one strategy for the interface is to strengthen the knowledge, and perhaps the role, of foresters. Foresters can be in the best position to resolve conflicts over forest values, especially those that involve timber harvesting. Foresters are trained professionals; they have the potential to see the various technical solutions that might resolve conflicting views. Foresters are usually familiar with the specific site and often know neighbors who will be affected by different land uses.

In recognition of this, California places strong emphasis on the role of professional foresters. To do forestry work, persons must be licensed by the state or be directly supervised by a person with a license. Licenses are granted based on experience and on passing a lengthy examination. Foresters who do not perform up to a professional standard are subject to discipline and may have their licenses suspended or revoked.

As licensed professionals, foresters in California can speak definitively about technical solutions to interface conflicts. But foresters in California generally have been hesitant to assume such a role. This is partly because most foresters work away from urban areas and are not involved in many controversial land use decisions. Other reasons are lack of training in public relations and a desire to avoid the formal political process.

A catalyst to increase the involvement of foresters could be the expansion of their knowledge of the politics of the interface. Henry Vaux has suggested several areas of possible improvement: better knowledge of techniques to influence politicians, a more flexible view of technological options, and 'greater understanding of and tolerance for conflicting views of what the social role of forests ought to be' (Vaux 1982, pp. 45-46).

In California, this is being accomplished through both continuing education and participation of foresters directly in the political process. Continuing education is provided by the Society of American Foresters and University Extension of the University of California. Political participation historically has been provided through work with the State Board of Forestry and the Department of Forestry. Recently, licensed foresters have organized into an association and have become an active political voice. This exposure to the political process has the dual benefit of making foresters more sensitive to urban political concerns and of making politicians more aware of the views of foresters.

Establishing a Policy Planning Mechanism

Sometimes issues emerge in the political process only after viewpoints have polarized. Controversial issues reflecting different values are then resolved by tests of political strength. Solutions derived from this process may not be the best, either technically or socially.

For this reason, it is important to have a way to anticipate policy issues and resolve them, if possible, before they reach heated political debate. California has, in part, developed this kind of mechanism. In 1977 the legislature adopted the Forest Resource Assessment and Policy Act. This law sets up a unit in the Department of Forestry charged with resource assessment and analysis. The unit works with resource survey programs of the U.S. Forest Service to provide data and analysis relevant to the state's needs. It also carries out an assessment of forest resources from the state's perspective. Part of this assessment defines critical issues to be considered by the State Board of Forestry. The Board, in turn, develops a forest policy statement for the legislature. This policy statement outlines key policy concerns and may suggest solutions. In this way, the legislature can anticipate and focus on issues before they emerge in the political process.

More work needs to be done in this area. Better ways have to be found to involve the public, at both the assessment level and the policy level. In this way, important elements of the public will sense that they are involved, and emerging issues can be defined more clearly.

DEVELOPING A METHODOLOGY TO IDENTIFY AND ADDRESS LAND USE IMPACTS BEFORE THEY OCCUR

One characteristic mentioned earlier about the urban/forest interface is that controversial issues have tended to focus on one piece of land. It is therefore important to identify and address site-specific impacts of various land uses before such uses occur. Good analysis, planning, and early involvement of the interested members of the public provide a basis for discussing differences over how the land should be managed. Potentially volatile political confrontations may thus be avoided or the intensity lessened.

An elaborate system of identifying and mitigating possible adverse environmental impacts of projects has evolved in California. Projects requiring governmental approval include subdivisions, construction of government buildings, and generally anything that can adversely affect the environment. Under the California Environmental Quality Act (CEQA), any proposed project must be reviewed for possible significant adverse environmental impacts. In some cases, this means that an environmental impact report must be prepared. The public is given an opportunity to review these documents before a decision is made on the project. Governmental agencies are obligated, by law, to consider and incorporate feasible mitigating measures.

CEQA has been in effect since 1970. A number of problems, such as prohibitive cost and time delays in receiving permits, have arisen and some have been addressed. Other problems remain, but CEQA has at least provided a framework to identify and discuss land use impacts.

CEQA also applies to timber harvesting operations, but in a modified form. Timber harvesting in California is subject to the Z'berg-Nejedly Forest Practice Act. Under this act, the State Board of Forestry makes forest practice rules. Rules are enforced by the Department of Forestry. Timber harvesting plans must be prepared by professional foresters licensed by the state. Plans generally contain information about how trees will be harvested, how streams and other resources will be protected, and how the land will be restocked with trees. Plans must be approved by the director of the Department of Forestry as being in conformance with the forest practice rules.

Timber harvesting is a project approved by the state and therefore subject to CEQA. However, because of the potential increase in complexity, cost, and uncertainty involved in CEQA's application to timber harvesting, special adaptations have been made. No environmental impact reports are necessary. Foresters are required to review proposed timber operations to see if they would have a significant adverse impact on long-term productivity or nontimber values enumerated in the Forest Practice Act. Consistent with forest practice rules, foresters must incorporate feasible mitigating measures to lessen environmental damage. Interdisciplinary review teams analyze all proposed timber harvesting plans, determine possible significant adverse effects, and suggest mitigating measures. The public is notified of proposed plans and has an opportunity to comment before the director makes a decision. The director considers these comments and formally responds to written public input.

In some cases, timber harvesting also has been subject to local regulation. Under the Forest Practice Act local and certain regional units of government have been able to regulate forest practices more strictly than the state. Five counties in the San Francisco Bay area and the Tahoe Regional Planning Agency have imposed additional regulations. These regulations can require performance bonds and usually control such things as traffic flows and hours of logging. The county ordinances also require a county permit in addition to a state permit. Sometimes these permits are subject to requirements for substantial additional environmental information. Permit requests are reviewed by county planning commissions and boards of supervisors, and sometimes involve much public input.

This local authority has been abused in a few cases. Heavy extra costs have been imposed on proposed harvesting plans in the form of extended delays and extra detailed environmental information. Consequently, new legislation was passed in 1982 that prohibits counties from adopting regulations or requiring permits. Beginning in 1983, counties can recommend regulations and the Board of Forestry must adopt them if they are consistent with the intent of the Forest Practice Act and are required by local needs and conditions. Local boards of supervisors may request public hearings on any proposed plan. It is likely that many of the same procedural requirements for public involvement now in place in these counties will be incorporated into new regulations.

This recent reaction in California to strong local regulation shows the need for understanding on all sides. Timber owners in urbanizing areas should expect more public concern and be prepared to respond appropriately. At the same time, ways must be found to anticipate and resolve opposition to timber harvesting expeditiously. If this is not done, substantial cost and uncertainty can be added for timber growers.

Generally, the strategy of identifying adverse environmental effects and planning mitigation for timber harvest operations seems to have worked well. Green et al. (1981) have found that procedural requirements of the Forest Practice Act and those developed in response to CEQA have been more influential in affecting operations than the rules themselves. They concluded: 'Requiring regular field inspections on all plans, and establishing sanctions for violations have played a part in bringing about a high degree of compliance with the standards, but the key to the situation has been the emphasis on procedures that anticipate problems before they occur, rather than punishing infractions afterwards' (p. 12).

KEY IMPACTS OF URBAN PRESSURE

The urban/forest interface reflects the pressures of a larger, more mobile population concentrated on a fixed forest land base. More people desire to live or to find recreation in the forest. Conflicts in forest use are more unavoidable. In addition to more intense political controversy, urban pressure has certain important effects on the forest land resource which must be considered if the resource is to be kept intact. These include higher land values, fragmented landownership, and the land use impacts of more people moving into forested areas.

Higher Land Values

A common effect of urban encroachment into resource lands is that land values rise greatly and no longer relate to the value of land in producing a resource such as timber or agricultural crops. Rather, values are related to the potential for things like homesites or recreational use.

Higher land values work against continued use of the forest resource for timber production and grazing. One example is that under an unmodified ad valorem property tax, higher property values lead to higher taxes; since timber growing renders only periodic income, higher annual tax payments make carrying costs prohibitive. Another example is that because forest land itself is worth more, local governments may become more dependent on tax revenue from such land and will resist efforts to hold down assessed values.

Control of rising land values is critical if society desires to keep resource industries like timber and agriculture in an economically viable condition. The main tools for such control are tax policy and zoning.

California, like a number of timber producing states, has modified the ad valorem property tax as applied to timber and timberland. This removes or lessens the financial effect of property taxes on management decisions. Similar to the state of Washington, California has enacted a percentage (yield) tax on the value of timber harvested. Timber is exempt from annual taxation. Timberland that is zoned as timber production zone (TPZ) in accordance with the yield tax law is valued based on its ability to grow trees. Land values are limited by law and by the use of strictly comparable sales data for assessment purposes. The valuation procedure is centralized to provide for uniform administration and application throughout the state.

In addition, California has developed a strong system of zoning lands for timber production. Subject to requests of the owner and statutory guidelines, land is zoned as TPZ by local government. Land in TPZ can be used only for timber production and compatible uses. Local governments determine compatible uses subject to a limited statutory definition of the term. The zoning is for a period of ten years. For every year after initial zoning that an owner does not petition the local government body to rezone his land, an additional year is added to the restriction. Thus, this ten-year zone is sometimes called a 'rolling zone.' Quick rezoning in less than ten years is possible, but it is difficult and very expensive.

Zoning land as TPZ helps control the influence of urban pressures on increasing land values. Land use is restricted, and speculative pressures are dampened because the zone runs for ten years. Available evidence shows that lands zoned as TPZ sell for less than lands not so zoned. This is some indication that within TPZ, land can be acquired at a price related to its ability to grow timber.

Forest lands not zoned TPZ can have higher values and be more subject to tax pressure. The effect has been moderated to some degree by general property tax relief, but higher values are still evident.

Despite TPZ zoning, it is still uncertain if land prices can be held down enough in the long run so that production of resources, like timber and cattle, can remain viable. Experience with agricultural lands in California, especially with highly productive lands, is that urban pressures eventually make resource production uneconomical. Perhaps the same will be true for forest land. Urban pressures that distort land values away from what the land can produce are so strong that they may overwhelm traditional approaches to limit these values. If this is the case, new strategies will have to be developed.

Fragmented Ownership

One effect of a strong urban influence is greater fragmentation of ownership of resource lands. This fragmentation happens geographically near urban areas and on forest lands within traveling distance of urban areas. It generally reflects people's preference to live and recreate in forested lands, as well as their ability to afford such land.

Increasing fragmentation of ownership is troublesome for several reasons. First, fragmented ownerships are often much smaller and less economical to manage than the larger ownerships from which they came. The land may not generate enough income to make it worth managing, so the ability to realize the productive potential of the land can be reduced or lost.

Second, land is often held by absentee owners. These owners have diverse, sometimes conflicting land use motives. They are difficult to involve in the local planning process, but represent a volatile political force if they become aroused.

Third, fragmentation may occur around larger parcels that remain economical to manage. Concerns of owners of small parcels over things like herbicides or clearcutting constrain management options on larger parcels. Finally, once the process of fragmentation has started it may be hard to stop. Substantial profit can be made from further land subdivision. Landownership may occur in smaller and smaller units. Development can be unplanned and not environmentally sound.

California has tried several tactics to deal with fragmentation. One approach has been to stabilize the forest land base by zoning land for timber production (TPZ). The ten-year period necessary for normal rezoning and the strict, expensive procedures for immediate rezoning discourage fragmentation.

The attempt to stabilize the forest land base has been relatively successful. Approximately 5.5 million acres have been zoned, or about 70 percent of the total private commercial forest land base. Almost no rezoning has occurred. This is deceptive, however, because not all forest land was zoned as TPZ by local government. Substantial pockets of non-TPZ land are intermingled with TPZ land. Fragmentation is still a potential problem in these areas.

A second approach has been to require local governments to plan for and evaluate parcels that split into more than four smaller parcels. Maps, called subdivision maps, must be prepared. Maps approved by local governments must be consistent with county general plans.

General plans are required as part of a substantial body of state law governing planning and land use. Every county and city must prepare comprehensive, long-term development plans for their land areas. General plans consist of a statement of development policies and related objectives and standards, including basic elements such as land use, circulation, housing, conservation, and open space. Plans must also address policies for land zoned for timber production (TPZ).

The intent is that basic land use decisions will be made in the general plan and implemented through consistent zoning. This provides local government with an opportunity to adopt a policy discouraging fragmentation. Such policies can be implemented by controlling the minimum parcel sizes allowed for land splits.

Policies differ with each local government. Counties that are more urban are somewhat restrictive of land divisions in forested areas. Rural counties tend to be less restrictive.

A third program developed in California, although not directly designed to deal with fragmentation, helps address the problem of small, less economical ownerships. The California Forest Improvement Program gives grants of 80 to 90 percent of forest improvement costs to persons owning less than 5,000 acres. Grants are made to finance land management planning, reforestation, thinning, habitat improvement, and erosion control work. To qualify, owners must have at least 5 acres that need improvement, but minimum acreage limits are not applicable to land conservation and habitat improvement projects. If the land is not zoned for timber production (TPZ), an owner must agree to keep the land in forest use for ten years.

So far, the program has been relatively successful. As of May 1982, a total of 352 projects have been authorized at a commitment of $3.4 million (California Department of Forestry 1982b). Over 50,000 acres, some of these in fragmented ownership, have been brought under management plans. Long-run viability of the program mostly depends on the level of sales from state forests and other funding sources.

The TPZ zoning and general plan requirements appear to have slowed fragmentation of some forest land. Current poor economic conditions have also diminished the rate of parcel splitting. However, over the long run, fragmentation is still a threat to the continued productivity of the forest land base. Outside of rigid zoning and strict controls on land use, perhaps the most promising strategy is to develop more programs like the Forest Improvement Program which will try to improve productivity on these small parcels. Improved productivity will enhance the economic return from keeping small parcels in resource production. This may offer an economical alternative to pay for the holding costs of land and offset financial pressures for subdivision. And, even if parcels continue to be fragmented, some portion of the land's productivity can be maintained.

More People Living in Forested Areas

One of the most significant effects of urban pressure is that more people choose to live in and use forested areas. This is especially true for forest land adjacent to urban areas or within traveling distance of them.

People arrive in the forest with attitudes to landownership formed in urban areas. They usually have little knowledge about forests. As more people concentrate in forested areas, there is increased risk of environmental damage from home construction, greater chance of wildfires starting and destroying property, and more pressure on water supplies and roads. Unless sufficiently controlled, such strains can ruin the qualities of forest land that make it productive for timber growing or attractive as living space.

The effects of urban pressures on California's forest lands are very apparent. The rural areas of the state are growing faster than urban centers. Growth is especially pronounced on the fringe of population concentrations in northern California. Two impacts of this growth are noteworthy: environmental damage from home construction and increased risk of wildfire.

Control of environmental damage from home construction is left primarily to local government. Subdivision planning, grading ordinances, and building permits are the principal methods of identifying and mitigating possible adverse effects on the forest resource. The strength of ordinances varies with each local government.

Occasionally, timber harvesting plans are submitted that contain roads designed for subdivision purposes and landings that will serve as housing pads. The Department of Forestry reviews these plans in the context of timber operations, not proposed subdivisions. Some environmental safeguards are thus provided, but again, effective control depends on the strength of local government.

As a general rule, architects, contractors, and others associated with development are unfamiliar with the impact of construction and landscaping on native vegetation and associated ecosystems. For example, improper sewage installation can cut into root zones of trees, killing the trees outright or making them vulnerable to insect attack. Similarly, lawns and exotic species of vegetation that require heavy watering may be introduced into ecosystems that contain vegetation that will not adapt to so much irrigation. Native tree species may die or be weakened. Practices that are insensitive to biological requirements of existing species can result in cumulative destruction of trees from insects and disease and other causes. Over a large area this can have significant adverse effects on watershed characteristics and cause increased erosion and flood hazards.

The strategy adopted to deal with this problem is a combination of education and specially designed local ordinances. Under the state's urban forestry program, ways are being designed to alert developers to special requirements of forest ecosystems. In addition, guidelines have been published for developing ordinances in urban forests. Key elements of these ordinances include provision of authority to carry out forestry programs, definition of responsibility for care of the urban forest, and establishment of minimum preconstruction, construction, and postconstruction standards (California Department of Forestry et al. 1982).

With regard to the problem of greater fire risk, the existence of more homes complicates fire control strategy and raises the risk of property loss. The potential for greater fire damage can be especially great if there is accumulated natural fuel.

If fire is a significant threat, a general strategy for an area might include programs for reduction of accumulated fuels, fire-safe design in homes and subdivisions, fire prevention and education, and ultimately an effective fire-fighting force.

All of these programs are found in California. Reduction of excessive fuel loads is handled through a chaparral management program. Cost sharing is offered to private landowners to use prescribed fire methods to burn chaparral. Liability for escaped fires has been largely removed.

In providing this program, the state has decided that it is cost effective to subsidize removal of fuel under controlled conditions rather than to incur the costs of fighting an uncontrolled fire.

Proper design of homes and subdivisions to fire safety standards is encouraged, but is not required by state law. Fire safety standards have been developed by the Department of Forestry, and a number of counties have incorporated them into local ordinances. Other counties have not adopted such ordinances, so there are still areas without adequate fire safety measures.

The failure of all counties to have such ordinances results from conflicting political forces. Efforts by the California legislature to control land use are resisted by local governments and special interest groups. New restrictions on local government are difficult to enact. This is especially true if state-mandated actions result in extra local costs, because the state is responsible for funding the implementation.

Given these forces, about all that can be done is to extinguish fires when they occur and to accept the greater level of risk to life and property. As long as fire insurance is available at affordable rates, it may be less expensive for property owners to pay insurance premiums than to incur the cost of making homes more fire safe.

Fire prevention and education historically have been important elements in California's program. Substantial amounts of literature have been distributed and many contacts with property owners have been made. The major drawback to the program is the inability to demonstrate that it is cost effective in preventing or reducing fires. As a result, budgets have been cut severely in recent years. Increased reliance is being made on organized citizen volunteers. Use of volunteers is a promising strategy for fire prevention in tight fiscal times.

With respect to fire suppression on forest land a rather elaborate system has evolved in California that emphasizes rapid fire suppression by an integrated, widely scattered fire-fighting force. The intent is to provide a level of protection that will minimize damage to the resource values on wild lands. Although it will respond to structural fires, the state does not provide primary protection to homes in forested areas. In lightly developed areas, protection for homes is provided by volunteer fire departments. As population density grows, people can choose to increase the level of structural protection. By agreement the protection can be provided by the state but paid for by local government.

California pays most of the bill for wild-land fire protection on areas of state responsibility, including most private forest lands. This reflects the judgment that the areas protected provide benefits to the public at large. It also works as a strategy to help support resource owners, such as timber growers and ranchers. State-funded protection operates as a subsidy; without the state, owners would have to develop and pay for fire protection on their lands. Some owners, especially timber companies, contribute substantial amounts of manpower and equipment to fire control efforts. But, even with this, Green et al. (1981) have estimated that forest protection support to timberland owners averages $3.20 per acre. This is a significant aid to such owners and helps offset some of the costs resulting from urban pressures on forest lands.

CONCLUSION

Intense and conflicting political views about the value of forests to society result from a larger, more urban population. Interface strategies are thus largely political. They focus on people's value perceptions, facilitate communication, and foster resolution of differing points of view.

At the same time, urban pressures unleash forces that must be held in check to preserve the basic forest land base. Critical factors include higher land values, fragmented ownerships, and adverse land use impacts of development projects.

In California the politics described by the interface and the strong pressures of population on forest land are very apparent. A number of strategies are in operation. Most have been developed only in the last few years. Since 1972, California has adopted licensing of foresters, a new Forest Practice Act, a different scheme of taxation for timber and timberland, a forest improvement program, an urban forestry program, and use of prescribed fire for chaparral management. The state has taken an aggressive approach to funding its renewable resource programs with such funds as the Forest Improvement Fund, the Renewable Resources Investment Fund, and the Energy and Resources Fund.

So much has happened in the last decade that it is difficult to assess the overall effectiveness of these strategies. As far as timber owners are concerned, Green et al. (p. 7) have estimated that the state, through all of its programs, subsidizes timber management by $3.80 to $4.50 per acre; costs of forest practice regulations range from $6.70 to $8.30 per acre, so net costs to timber owners range from $2.90 to $4.80 per acre. No comparable estimates have been made for other aspects of state programs.

Only time will tell if California's aggressive approaches will work. In recent years forestry issues have been hotly debated at both local and state levels. Differences in values about how forests should be used run very deep, and some political conflicts are inevitable.

In such climate it is easy to lose sight of what is happening to the basic forest resource. Both private and public monies are needed to keep forest resources productive. Political differences must be set aside long enough to focus on problems of promoting sustained investment in forests. Perhaps this is the real starting point in discussing the urban/forest interface.

REFERENCES

California Department of Forestry. 1982a. Annual report, 1978-80. Sacramento, California.

———. 1982b. Resource management, monthly activity report. Sacramento, California. June.

California Department of Forestry, U.S. Forest Service, and Cooperative Extension, University of California. Guidelines for developing ordinances in urban forests. Sacramento, California. June.

California State Board of Forestry. 1981. Renewable resources under siege: A policy for California forestry. Sacramento, California. October.

DeBruin, H. W. 1982. Urban and community forestry: The time is now. American Forests 88(3):19-25, 52.

Green, P. F., F. L. Norbury, M. N. Palley, and H. J. Vaux. 1981. Regulation of forest practices in California. Department of Forestry and Resource Management, University of California, Berkeley.

Vaux, H. J. 1982. Forestry's hotseat: The urban/forest interface. American Forests 88(5):37, 44-46.

20

STRATEGIES TO ACHIEVE PUBLIC AND PRIVATE LAND USE AND FOREST RESOURCE GOALS

Donald H. Miller

In addressing this topic, I will emphasize an urban view—a view from the city and suburb toward the forests. Thus I will review some of the values discussed earlier in this symposium that are served by conserving forested areas, but focus on those to which urban residents give highest priority. These become the purposes urban residents have for retaining forested lands in urbanizing areas.

Most urban residents like trees. I will not go into the reasons for this, but this interest is reflected in street tree programs, urban parks and forests, the addition to market value of a tree-filled landscape treatment of residential and other urban developments, and increased attention to tree preservation in modern subdivision ordinances and zoning provisions, such as those dealing with cluster development and planned unit developments. Although trees in these settings are the most important to urban residents, because they are the trees most accessible to them, my focus will be conserving forests in urbanizing areas—the larger contiguous wooded areas that could or do have value for growing fiber, including industrial forests.

What are the dimensions of the forest land conversion problem? Various sources differ, but as of 1977, nonfederal forest land encompassed 377 million acres, or 25 percent of the nonfederal land area of the fifty states (Dederiksen et al. 1977). Urban land accounted for 90 million acres, or 6 percent of this land area. However, according to USDA data, nonfederal land occupied by forests decreased by 31 percent during the period 1967-77, while urban land increased by 68 percent. Clearly, not all conversion of forest land is due to urban development, and some estimate that only half of the land classified as commercial forest land is currently producing timber. But recent conversion of forest land to other uses is cause for concern.

Forests with commercial value are threatened by urban development in at least three major ways. The most obvious is the spread of urbanization: conversion of land at the edges of urbanizing areas to urban uses such as homes and commercial and industrial activities.

The second and slightly less obvious source of conversion is the development of homesites in nonurban settings: the development of recreational second homes, homes to which people retire, and recently homes in rural locations for families fleeing city life and often trying to make subsistence use of their property. Some colleagues and I are beginning to look at still another form of 'footloose' residential location, based on emerging computer and communication technology permitting people to work at home, which could greatly increase demand for homesite development in rural locations and thus further affect commercial forestry. These developments are not contiguous to existing urban areas and, while often rural in character, are essentially urban in use (Beale 1976). There is evidence that these developments, when they are neighbors of commercial forest tracts, can so constrain forestry practices, and increase tax assessments and rates, that they affect commercial forestry perhaps even more than does the spread of existing urban areas.

A third major way that urban development contributes to the conversion of forest lands is even less direct. About 520,000 acres of rural land are converted annually to nonurban but largely urban-related uses, including transportation, reservoirs, strip mining, pipelines, power lines, and other energy-related uses (Esseks 1979). While much of this land conversion is at the expense of forested areas, I will not emphasize this class of land conversion.

URBAN PURPOSES FOR RETAINING FOREST LANDS

With these comments as a framework, let us consider the values or purposes that urban residents give highest priority to in retaining forests, especially in urbanizing areas. The primary purpose is aesthetic: to provide open spaces around or within sight of one's residence or place of work. Although most residents do not think about the contribution of trees to the visual landscape as analytically as planners and other designers do, stands of trees that define views, accentuate land forms such as hills and stream valleys, and help to screen different land uses are especially valued. While some of the wooded areas that serve this purpose are committed to landscape uses or are too small to be of much value for fiber production, substantial amounts of commercial forest land in urbanizing areas are valued for this purpose.

This is not only the case with forests visible from urban developed areas but, as some forest owners are painfully aware, it is also the case with forests visible from highways and rural roads. Visible clearcuts have become a major political issue in several parts of the country over the last years.

Aesthetic contributions produce little or no profit for the commercial forest owner. In fact, they tend to have a spillover effect as far as the forest owner is concerned: they represent a value received by others without compensation to the property owner.

Environmental protection is also important to urban residents for retaining forested areas around cities. Environmental values include watershed protection, land stabilization, flood control, groundwater recharge, and air quality protection or improvement. These positive externalities sometimes are so closely related to health, safety, welfare, or other public purposes that they are clearly subject to exercise of police powers, as through zoning and other regulation.

Next in priority is recreation, in the form of opportunities to walk, camp, and hunt in forested areas, and more recently to use all-terrain vehicles and snowmobiles. These activities require securing agreement from the private owner or some form of public ownership of the forest. In short, they exceed the sorts of externalities that provide a basis for regulation of land.

A somewhat related purpose is wildlife conservation, not only to retain habitats for populations of indigenous flora and fauna, but in rarer cases to preserve critical biological areas. The latter involves preservation of an area and so limits land use that conservation easements or acquisition of the property is required.

Close to the bottom of the urban resident's list of purposes for retaining forests is fiber production. While urban markets are clearly the major consumers of wood and wood products, there is little awareness of where this fiber comes from, of possible future supply problems, or of the economic importance of forestry and the wood products industry—which is major in the case of Washington State—to the region or to the locality. This lack of awareness and concern could change, as we are seeing in some areas regarding production of agricultural goods close to the urban markets, but such a development will require a substantial educational effort.

This is a brief summary, and rough priority, of the purposes that motivate urban residents to want to retain forests in urban areas. These values and their ordering are shared by many nonindustrial forest landowners. Several of these purposes are complementary. But industrial forest landowners would for the most part reverse this list. This, I think it safe to say, is a major basis for conflict between urban residents and owners of commercial forest lands. Recognition of this difference in priorities can provide us with a start at resolving this conflict, through mutual education and compromise.

Any effort to address the problem of retaining forests in urban areas must also deal with two major social goals, both of which tend to be more important than the purposes just summarized: (1) to supply adequate land for housing and other urban uses (this is often required to avoid discriminatory results, and can have a substantial impact on land costs and, consequently, housing costs); (2) to respect legally specified rights in land.

LANDOWNERS AND THE LAND MARKET IN URBAN FRINGE AREAS

Over roughly the last decade, those concerned with understanding the effect of various programs to influence land use and development in the urban fringe areas have come to recognize the importance of understanding the dynamics of the land markets in these areas (McBride and Clawson 1970, Kaiser and Weiss 1970, Brown et al. 1981, Healy and Short 1981). Therefore it is necessary to identify the owners of open land that is feeling or is likely to feel the pressure for urban development, and to identify the motives and behavior of these owners—their reasons for selling, for making locational choices, and so forth. Only then can we begin to assess how responsive these owners, as well as sellers and buyers, will be to various controls and incentives available to implement public policy regarding retention of forested lands.

This knowledge of market dynamics is critical not only in assessing the effectiveness of strategies for conserving wooded areas but also because it is through the responses of owners and other actors in the land market to these public programs that the differences between public and private goals get resolved.

As yet, our knowledge of these market dynamics is incomplete. The general conclusions of empirical work over the last few years are summarized by the statement that 'because of the opportunity for speculative profits from appreciation in land values, the character of both rural land and its ownership begin to change more than twenty years before an area is actually urbanized, and long before public policy efforts to influence urban development typically take form' (Brown et al. 1981, p. 131).

First, who are the owners of urban fringe lands? These owners may be typified as either 'land users' or 'land holders' (Lindeman 1976). Land users regard their property as a factor of production, and value it accordingly. Users include owners who grow food or fiber, reside on or otherwise occupy the property, and developers buying at the time of intended new use. Land holders own property in speculation for future profit due to anticipated increase in value. Land holders are often referred to as investors or speculators.

When the market presents the possibility of converting property to urban development, land users often become speculators as well. Owners begin to regard their prior use as impermanent, lowering their long-term investments (Miner 1976, Berry and Plaut 1978). This is a major spillover or indirect effect of nearby urbanization. For example, it is estimated that in sections of the Northeast, one acre of rural land is idled because of urban pressures for every acre converted to urban uses (Plaut 1976).

Small Private Landowners

Most of the nonindustrial forest landowners and farm landowners in the urban fringe areas are small private landowners. These owners use their land as a productive resource, but it is among the most readily available for purchase and conversion. Consequently, understanding their motives and behavior is critical to formulating effective strategies for keeping their land in rural uses.

Investigating four U.S. and two Canadian metropolitan areas, Brown and colleagues (1981) divided the undeveloped land into land likely to be developed within ten years ('areas of intense development pressure') and land not likely to be developed for at least ten years and perhaps longer ('areas of moderate or weak development pressure'). In the zone of moderate or weak development pressure, 84.5 percent of the land was owned by individuals or families, 82.1 percent was acquired for personal use, and owners of 81.2 percent expected to continue the existing use and to own land for at least five years. Proportional uses of this land were 48.3 percent in agriculture (38.3 percent responded that they farmed the land personally), 32.7 percent in nonagricultural use, and 19.0 percent not in use. About a third of the land in this zone was held in parcels larger than 200 acres.

Comparison of these and other data with their counterparts for the zone under intense development pressure points up the importance of ownership by families that are land users. Landowners most often cited personal factors (36.4 percent) such as changes in family composition and their own advancing age as the reason for selling their land, suggesting that demographic rather than economic factors were responsible for land in these study areas going on the market.

Of reasons given by city residents for owning forest land in western Oregon, 38 percent cited the satisfaction of owning land, 20 percent the satisfaction of working their own woods, 26 percent the income from timber sales, and 38 percent the investment for resale. While 80 percent of the owners of nonindustrial private forest lands in this area hold parcels of less than 100 acres, 53 percent of this land is held by 5 percent of the owners in parcels over 500 acres (Downing et al. 1976).

A recent study in Washington looked at a different set of forest owners: those belonging to the Farm Forestry Association or with property designated as timber tax land. Only 14 percent of the parcels of these owners were under 20 acres; more than half reported that their reason for ownership is to grow and harvest timber; and about 40 percent intend to own their land for twenty-five years or longer (Koss and Scott 1978).

Although the data are not conclusive, evidence that we do have suggests that nonindustrial forest owners, like other rural land users, have 'significant nonpecuniary motives in addition to their pecuniary motives' for owning land, and that 'these intangible motives increase the present value to the landowner but do not affect market value' (Kaiser and Weiss 1970). This has suggested to Gould that we should speak of 'high satisfaction' instead of 'high yield' forestry in these cases (Gould 1978). These motives also suggest how owners will feel about selling their land. Key factors in identifying those owners least likely to sell their land were noted by one study: living on the site, nonretired, own singly, and have held the land for at least ten but not more than forty years (Kaiser and Weiss 1970). Various research results tend to corroborate this, leading to the conclusion that a high proportion of nonindustrial forest owners in areas with moderate or weak pressures for urban development wish to hold on to their land, and many of those in areas with more intense conversion pressures wish to do the same.

Industrial Forest Owners

Industrial forest owners are clearly land users, with less complex motives than nonindustrial owners. These motives emphasize managing their holdings in a manner that maximizes timber yield, translated into profitability. They usually hold very large land parcels and, as in many of the counties in western Washington, hold a high proportion of the existing rural land. Yet their motives and behavior are not simple. They appreciate environmental considerations and outdoor recreational opportunities, and often incorporate these into multiple-use management practices (Clawson 1972). However, as businesses, often with stockholders, they are encouraged to act in a manner that gives priority to profit.

Industrial owners of forests in or near the urban fringes are thus ambivalent about these holdings: they face at least two conflicting objectives. On the one hand, they desire that tax assessors view their lands as timber producing and tax them accordingly. They consequently marshal arguments to support the position that this is a permanent use and important in providing materials for their processing facilities. On the other hand, they often appear to want to avoid restrictions on converting their holdings in close-in locations to urban uses, such as sale for homesites or recreational development.

In short, industrial owners are land users with a long-range time horizon, who also want to retain an investor or 'land holder' option. Sales of the land, or even direct development through a real estate arm of the firm, tends to take place when their reserve price is met, and especially when the resource use value of the land declines because of surrounding development.

Investors

Land holders or investors are the last group of owners that I want to mention. Investors acquire land mainly to realize a profit through resale after an increase in market value results from increases in demand and the indirect effects of investments made by others. Land purchased by investors is almost by definition converted to a transitional use. Investors seek to minimize their own payments by using substantial loans for their purchases, are concerned about interest rates and other costs of holding the land, commonly seek resale in a relatively short time, and are attentive to the income tax implications of their transactions (Lindeman 1976). Their motive for ownership is economic, in contrast to the more complex motives of various other groups of land users.

Land may change hands in speculative purchases several times before being sold to a developer, with later investors being less sophisticated than earlier buyers. Financing for investment adds to the price of land held for speculation, encouraging developers to circumvent, or 'leapfrog,' these parcels in favor of land available for high-equity owners at less expense.

In the Brown study of the urban fringe lands of four U.S. metropolitan areas, 20 percent of the landowners were investors, holding 21 percent of the parcels and 18 percent of the land (Brown et al. 1981). When we look at these urban fringes in terms of development pressures, we can conclude that close-in areas are already largely converted to transitional uses. More than 30 percent of the land in this zone was owned by investors. In the areas offering greater possibility of conservation of forest lands (the zone of moderate to weak development pressure), only 14 percent of the land was owned by investors, although 24 percent of the land held for ten years or less was owned by investors. These figures illustrate that speculators are an important minority among owners of rural land in urban fringe areas. Consequently, it is important to consider the effects on this group of land holders of various public programs to retain forest lands, as well as the effects on those whose purpose in ownership is to use the land.

MAJOR MEANS OF CONSERVING FOREST LANDS

Most of the common strategies employed in the United States to achieve public and private land use and forest resource goals are public programs to limit conversion of forest lands to urban uses. These programs are primarily land use planning, regulation, and other activities at the state and local levels, especially at the county level. After a brief description of these activities, we will assess their strengths and shortcomings, and finally draw some conclusion concerning which strategies appear to be most promising.

Public programs to conserve forest lands, and to minimize conflicts between urban and forest land uses, fall into three categories: regulations, incentives, and public acquisitions. Regulations attempt to limit the use of certain land to forestry and other open and rural uses, as well as to low density residential uses, and to control permitted land uses so that the effects of these uses do not cross property or zone boundaries. Incentives are offered to encourage forest owners and farmers to keep their lands in production and open uses. Public acquisition involves obtaining property or limited rights in property in order to hold it in forested use and at times to gain public uses of the land.

Regulation

In the regulation of land uses and development, we are most familiar with the exercise of police power through such means as zoning and subdivision ordinances. Efforts to keep land in forest use defy clear classification along these lines, and major examples of programs to designate forestry as the exclusive use employ tax relief rather than exercise of police power. Most rural zoning deals with farm lands, but we have learned some lessons that can also be applied to zoning to conserve forested areas. Prominent zoning efforts to control conversion of rural land are linked to the definition of urban growth boundaries.

Forest and Other Rural Use Zoning. Several states and counties have instituted zoning regulations that make forestry and agriculture the principal or exclusive permitted uses of designated land. The Oregon Land Use Act of 1973, for example, requires cities and counties to zone areas that have highly productive soils, and are not needed for predicted urban growth, for 'exclusive farm use.' The plans and zoning of these local jurisdictions are reviewed and approved by the state.

California has instituted a timber production zone (TPZ) in which areas so designated by counties are to be used only for growing and harvesting timber and for related activities. While a major feature of this program is tax relief, exclusive forest use of these areas is binding. Owners may request that their properties be removed from this designation, but permission requires a four-fifths vote of approval by the County Board of Supervisors, after public hearing, and the land must remain in forest use for ten years after approval is granted.

Most efforts to apply police power through zoning to retain forested and agricultural areas have used minimum-lot-size zoning (Toner 1978). The reasoning is that large lots preclude use for strictly residential purposes and encourage food and fiber production on the land (Babcock 1969). Minimum parcel sizes of agricultural zones in California counties range from 640 to 60 acres. California's timber production zone allowed small parcels to be included initially, but established a 160-acre minimum size thereafter. A Washington county has specified 20-acre minimum parcel sizes for lands designated for commercial forest production use. These lot sizes are or should be based on the economics of the rural use of the land.

Most minimum lot sizes for agricultural zones established by counties under Oregon's Land

Use Act are from 20 to 40 acres. The higher purchase price of larger minimum parcel sizes has shifted demand for hobby farms to land zoned for rural residential uses, which have smaller parcel sizes (Gustafson et al. 1982).

Whether use and parcel size zoning for retaining land in food and fiber production will withstand pressure for urban development, or even rural residential development, is still open to question (Toner 1978). There is some evidence that where development pressure is not intense, and where there seems to be governmental resolve to uphold policy, investors and developers chose sites with fewer restrictions (Kaiser and Weiss 1970, Berry and Plaut 1978).

Most efforts to zone areas for rural uses have been based on the productive capacity of the land. This often means that some forest lands that would serve those amenity and other purposes given highest priority by urban residents are not included in these rural zones.

Large-lot zoning in recognized transitional areas has also been used. While the intention is to maintain low densities until urban demand requires higher densities, this practice has been faulted as contributing to expectations of residents that their area will remain in low density development, and as giving rise to inefficiencies when this land is redeveloped.

Yet from an urban perspective, selected lands retained for the long term in low density development can provide aesthetic values such as visually important stands of trees and other benefits such as runoff control and wildlife habitat. Accomplishing this through zoning is difficult if the ordinance is challenged, because of problems of avoiding the 'takings' issue and charges that the action is arbitrary. In most cases, forested lands conserved in this manner would not make substantial contributions to timber production.

Planned Unit Development and Performance Zoning. Traditional zoning has come to be recognized as excessively rigid, especially when specific provisions of an ordinance are applied to a variety of situations (CEQ 1974). Responses have included the increased use of planned unit development and performance zoning provisions. Both of these techniques have allowed for more flexible development of parcels of land, to encourage better design and use of the site, rather than dealing with the location and specification of uses. For our purposes, they are worth including as means of controlling development adjacent to commercial forest lands, so that the design of the development minimizes impacts between the adjacent uses.

Planned unit development provisions relieve a developer from strict setback, lot size, and some other requirements of the zoning ordinance, but not the permitted land use provisions (So 1973). With a freer hand in developing a parcel, the developer may cluster the housing, for example, often saving on utility and other site preparation costs, providing more usable open space, taking advantage of natural features, and setting the development back from property lines shared with other uses such as commercial forestry. Often an additional incentive to use this provision is the availability of a density bonus, allowing more units to be placed on the site than under the standard provisions of the ordinance. A minimum parcel size is usually required before this provision can be used by an owner, and the design for the development must be approved before the permit is granted.

Performance zoning is similar. Some of the standard provisions are relaxed if the proposed design for development scores well with respect to a set of criteria. Performance zoning has been applied in rural use zones, permitting the number of buildable lots to vary inversely with the size of the parcel and requiring that these lots be located on less productive soils. Tulare County in California employs a rating scheme for soils, surrounding uses, and several other factors to evaluate any application for a zoning change from rural use (Toner 1978). A variant of this is conditional or contract zoning, which allows flexibility from standard provisions if specified conditions are met and the proposal is found to be more desirable than if the standard provisions were applied (Hagman 1973).

Several of these approaches were designed primarily for urban settings. The very flexibility they intend to provide can give rise to problems. The permit process can be complex and time consuming for the developer, and neighbors may resist proposals that are inconsistent with surrounding development.

Growth Boundaries and Other Growth Controls. Many local jurisdictions have become concerned about the extent and location of urban growth, in addition to the quality and spatial organization of development. Over the last decade or so, local governments have increasingly turned to a physical planning and management system that addresses these concerns, stimulated in part by the Ramapo, New York, ordinance.

Most localities have recognized that a growth moratorium is legally and politically difficult, and have focused instead on channeling the rate and location of growth more efficiently. Many growth control programs are based on the recognition that, from a fiscal point of view, scattered development is more expensive than contiguous development (Real Estate Research Corporation 1974).

A major means for controlling growth involves defining boundaries within which urban growth is permitted, and implementing this through zoning, the subdivision ordinance, and the provision of basic urban services and amenities, such as sewers and water, roads, parks, and fire stations (Gleeson et al. 1975). When timing of public facilities is specified by a phased capital improvement program, development must take place within designated areas or else these facilities must be privately constructed to community specifications.

Examples of growth management programs include planning and implementation under the Oregon Use Act, and an effort currently under way in King County, Washington. In these cases, lands contiguous to existing urban development are designated as available for urban development in quantities adequate to supply predicted demand over a period of years. Locational choices for these development areas are made on the basis of the comparative costs of providing the public infrastructure, and encourage infill of areas already served by utilities. These programs also encourage areas of higher density development than are sometimes found in suburbs, as well as provision of lower cost housing.

There is evidence that strategies of this sort influence development decisions (Kaiser and Weiss 1970), and may be among the most effective means of avoiding conversion of nearby lands in rural use by encouraging development of areas that are most suitable, and in a compact form (Brown et al. 1981).

Timber Harvesting Ordinances. Several counties have passed timber harvesting ordinances that regulate nuisances and hazards that can be associated with harvesting practices, seek a balance between the commercial value of forests and the residential value of woods, and yet sustain timber production as part of the local economic base (Thurow et al. 1975). In addition to regulating firebreaks, log hauling, and noise, Napa and San Mateo counties in California require selective cutting procedures to prevent erosion and preserve surface water quality, and encourage leaving visually important stands of trees. It is pointed out that by minimizing conflicts, the local government helps maintain the viability of commercial forestry in an urbanizing area, and holds all timber operators to the same standard of behavior.

Some of these provisions are included in, for example, the Washington Forest Practices Act, but the state attorney general has given the opinion that counties may also regulate forest practices such as selective cutting, leaving buffer strips, and restricting plat approval on land recently harvested unless specified techniques had been employed (Washington State Association of Prosecuting Attorneys 1980).

Incentives

Incentives to forest landowners take two—often related—forms: (1) differential tax assessment of real forest property, and (2) formation of forestry districts, within which timber growing and harvesting predominate, employing tax relief as a major incentive.

Preferential Assessment. Preferential taxation is a means of decreasing the tax burden on forest and other rural land uses that a jurisdiction wants to retain, by assessing this land at less than full market value. Most states have adopted preferential assessment programs, reasoning that current tax policies discriminate against owners of land in open use, since the assessment is on the potential value rather than use value (CEQ 1974). Problems arise when properties not receiving preferential treatment are unfairly taxed to supply needed governmental revenues, and when speculators benefit from lower property taxes while they hold the land for appreciation.

There are three major forms of preferential taxation. The first is simple preferential assessment, whereby the land is taxed on the basis of its current use value. The second is tax deferral, whereby the land is taxed at its current use value, but if it is taken out of its current use, back taxes on the difference between its current use and market value, for a stated number of years, must be paid. A third and more demanding form requires an agreement between the owner and a governmental agency that the owner will continue the current use for a stated number of years in return for preferential assessment (Hady and Sibold 1974). For reasons of equity, and to guard against misuse, the two latter forms are usually employed. Two tax programs in Washington and one in California illustrate these forms of taxation.

Washington's Timber and Forest Lands Tax Act of 1971 makes it possible for forest landowners with parcels of 20 acres or more to obtain property taxation on their land based on current use value, and to pay a severance tax of 6.5 percent on their timber when it is harvested. The land value for property taxes is set by the state, and the owner must provide the state with a management plan. Removal of land from the program requires the owner to pay back taxes for ten years on the difference between the current use and market values of the land.

The Washington Open Space Law for Timber, also passed in 1971, allows owners with forested parcels as small as 5 acres to obtain about a two-thirds reduction in the property tax on their land if they allow 'reasonable public access' to their land. Approval must be secured from the county, which requires a management plan. Removal of land from the program requires recapture of back taxes for seven years on the difference between use and market value assessments, plus interest, and a 20 percent penalty.

The California Forest Taxation Reform Act of 1976 is an example of the contractual form of differential tax program. This act established a timber production zone that county assessors applied to lands taxed as forest use and other forested lands that owners wished to have included (Whitney 1977). As discussed earlier, removal of land from this zone requires action by the county legislative body, and the land must remain in the zone for ten years after the request for removal is approved.

The effectiveness of preferential taxation is difficult to assess. There is always concern, especially on the part of tax assessors, whether the owner is primarily interested in growing timber or in lowering out-of-pocket expenses while speculating on the land. Among motives of different groups of owners, economic factors including taxes were not found to be the most important variables influencing sales by private owner-users of the land, but were much more important to investors in affecting their holding costs.

Current use taxation does not facilitate purchase of forest land to continue growing timber on it, since the market price reflects the future urban use, which is greater than its current use

(Clawson 1962). Yet Brown found that 'well targeted' preferential taxation would influence 17 percent of the U.S. owners that he surveyed who were selling their land, and up to one-third of those owning land in the zone under intense development pressure (Brown et al. 1981).

Forestry Districts. While most of our experience in employing special-use districts that provide multiple incentives has dealt with agricultural uses, this strategy holds promise for forestry as well. California's timber production zone is a partial illustration. Oregon's exclusive farm zone is another.

New York's Agricultural Districting Program is an example of combining several protections, including the availability of preferential taxation with recapture provisions (Bryant and Conklin 1975). Farmers may voluntarily form a district, in which various farming practices may be carried out, and protection is provided against special assessments for utility districts and from eminent domain without careful consideration of alternatives sites for the public use. Such districts are created for fixed periods, with the option of renewal (Coughlin et al. 1981).

Public Acquisition

Purchase of land, or development rights in land, can be an important means of retaining forested areas valued highly by urban residents, especially where these areas are likely to be most subject to urban development pressures and where regulations and incentives fail (Stanford Environmental Law Society 1973). Several devices are available, which vary in cost and sometimes legality.

Fee simple acquisition is the most expensive, and involves full compensation for full title to property, commonly for a public purpose such as a park. Increases in nearby land values may provide new tax revenues which offset some of the cost. *Land banking* usually involves public purchase of property where urban expansion is likely to occur, and later resale or lease at urban prices, with deed restrictions. Although primarily employed in European countries, this approach has been used in Boulder County, Colorado, and Montgomery County, Maryland. Citizens in Lincoln, Massachusetts, formed a nonprofit foundation to purchase an important parcel of land, set about half of it aside as open space, and developed the remainder in a manner that they thought appropriate to the area (Lamire 1979).

The purpose of land banking is to control the timing, nature, and location of urban growth, although purchase and lease-back or sale of forest lands for continued forest use is a possibility. A public land bank agency would need adequate financing capabilities, and the power to purchase and hold land and to condemn property. If part of a program with public purposes, use of eminent domain would probably be legally acceptable (Gleeson et al. 1975).

Less than fee simple acquisition involves purchase of the rights to develop land, or a conservation or scenic easement. The landowner retains rights to certain uses of the land as well as the expense of managing the land. If the owner grants an easement at less than its market value, the extent of the contribution is tax deductible. The new market value, after development rights are severed, will mean that property taxes are reduced (Coughlin and Plaut 1978).

The financial burden of a program to purchase development rights is substantial, but King County has passed a bond issue to conserve some of its agricultural lands, and Wisconsin and New Jersey have developed funding for this purpose. The discovery that development rights can cost up to 90 percent of the market value of the land has suggested to some that fee simple purchase may be preferable.

While public sector funds for buying real estate to preserve important ecological or aesthetic values have been dwindling over the last few years, purchases of land or rights in land in the open market may bc thc most direct, simplest, least controversial, and even least expensive way of preserving key parcels (Fuguitt and Voss 1979).

HOW SUCCESSFUL ARE THESE COMMON STRATEGIES?

We have identified and discussed most of the major strategies employed by governments in this country to retain forest lands in urban fringe areas, with some emphasis on commercial forests. The motives and behavior of major groups owning these lands have been discussed to see how various public programs of regulation, incentives, and acquisition affect their selling and buying of forest lands. How successful are these strategies? One measure of success is the extent to which public strategies have prevented conversion of forest lands. We have insufficient data to test the effectiveness of strategies that are in place, and must infer answers from information concerning such things as motives and behavior.

A second means of assessing success is to ask how feasible or workable these strategies are. This question is as important as asking about effectiveness, because it helps us identify the problems of using these strategies.

Finally, do these strategies work better together than individually, and are they more useful in one part of the urban fringe than another? Let us first try to answer these questions for each of the three major types of public programs discussed earlier, and then for combinations of these programs.

Regulation

While direct regulation of land use would appear to be the most effective means of retaining forest lands, there is continuing concern that political or economic arguments will cause land designated for open uses to be rezoned or other controls to be relaxed. There is also concern that efforts to retain open rural uses through application of police power will be found unacceptable by the courts, although some argue that this problem is more political than constitutional (Gustafson et al. 1982).

The question of how much others may benefit through the spillover effects of requiring an owner to retain land in open use, while the owner is prevented from making financial gain by converting the land, is still unresolved. Recent court decisions have focused on whether current use value has been reduced through regulation, and whether the continuing permitted use has reasonable economic value.

Furthermore, while regulatory programs such as zoning for rural uses specify what land is to be included, the common procedure in adopting such an ordinance is to accommodate current owners when drawing the map, to avoid controversy or in response to political pressure. Consequently, some lands that would be desirable, and may even be the key for aesthetic and other purposes, may not be regulated to prevent conversion.

Controlling extension of public utilities and facilities through use of growth boundaries and a phased capital improvement program raises fewer problems of these sorts then does zoning. Yet this form of development regulation is subject to political pressures as well, and is often complicated by the presence of independent special utility districts. Continued political support of this strategy may depend on public awareness of the taxes saved by timing and channeling growth.

Incentives

Incentives to forest landowners, in the form of protection from the indirect effects of urbanization, but especially through preferential taxation, would appear to play a minor role in the decision to sell property. Personal and other factors weigh more heavily, as do other economic conditions such as the market value of the product from the land.

Tax relief would need to be very great to offset market value increases in areas of intense pressure for urban development, but more realistic tax relief in these areas can complement the effectiveness of direct regulation. Current use assessment in areas with weak development pressures probably has little effect on the decision to sell, except for properties of marginal productivity.

Of course, tax relief for some tends to mean that others must shoulder more of the tax burden, unless reduced services to preferentially taxed lands make up the difference. The amount of increased taxes and the benefits to these parties from lands retained in open use have yet to be measured (Gustafson and Wallace 1975).

Assessors have been outspoken about how difficult it is to determine whether an owner wants to grow timber or wants to take advantage of a tax break for another use or for speculation. Assessors are also troubled by the problem of determining current use value, when their practice is to use comparable sales.

Acquisition

Land and development rights purchased by the public, especially in areas under pressure for urban development, are very expensive. In addition to the problem of estimating the value of development rights, prices of these rights represent a high percentage of the market value, which usually means that only a small portion of the area desired for continued open use can be purchased. But this strategy can complement efforts to retain open uses through regulation, since zoning, for example, is of limited use where development pressures are intense.

CONCLUSIONS

This brief critical assessment of the relative effectiveness and problems with our major strategies for retaining forested lands leads to a major conclusion: none of these strategies are adequate by themselves, but they do complement each other in a number of cases. No one combination of strategies is likely to work in all situations. The nature and extent of development, patterns of ownership, local governmental systems, political values, enabling legislation, and more—all influence the appropriate design for accomplishing both public and private goals in controlling land uses and management practices in urban fringe areas.

In areas of the urban fringe that are affected by intense urban development pressure, regulation in the form of exclusive use zoning and performance zoning, controlling utility extensions, and other capital improvements, along with timber harvesting controls, stands the greatest chance of retaining lands in commercial forestry. Still, some key properties will be kept in open use only through public acquisition programs. Preferential taxation is of modest effect alone in this situation, but can partly offset the indirect effects of surrounding urbanization on forest management practices.

In portions of the urban fringe where development pressures are moderate or weak, zoning may be unnecessary to discourage speculation and will likely be opposed by property owners.

The exception to this will occur where development of second and rural homes threatens to convert parts of an area in commercial forestry. Zoning offering protection from this development, or from the indirect effects of this development when it occurs, could be of economic value to forest management on adjacent sites. This type of protection could also be provided by forestry districts.

This approach of combining public programs of various kinds is the intention but not always the practice in growth management programs. The evidence suggests that communities *have* access to adequate strategies for shaping urban patterns. 'It is a question of how, when, and under what conditions these controls are exercised' (McBride and Clawson 1970). By exercising foresight and political will, communities can have considerable control over their own destinies.

A second major conclusion is analogous, I think, to the old saw about paying more attention to the hole than to the doughnut. That is, the greatest effectiveness will be achieved by focusing not solely on retaining open land but also on accommodating growth and channeling it to the most appropriate sites in compact form.

We need to plan for urban development, not simply plan to prohibit urban development of some areas. We need to emphasize what is desired rather than what is not desired. We need to make adequate and attractive space available for urban activities in a manner that encourages developers to locate there, and to deflate development expectations for land planned for forest and other rural uses. Finally, sound comprehensive planning and policy is not only environmentally and fiscally attractive, but it is important in surviving legal challenges to the strategies that we use to conserve open uses in our urban fringe.

REFERENCES

Babcock, R. F. 1969. The zoning game: Municipal practices and policies. University of Wisconsin Press, Madison.

Beale, C. L. 1975. The revival of population growth in nonmetropolitan America. Economics Research Service. Publication 605. U.S. Department of Agriculture, Washington, D.C.

Berry, D., and T. Plaut. 1978. Retaining agricultural activities under urban pressures: A review of land use conflicts and policies. Policy Sciences 9(2):153-178.

Brown, H. J., R. S. Phillips, and N. A. Roberts. 1981. Land markets at the urban fringe: New insights for policy makers. Journal of the American Planning Association 47(2):131-144.

Bryant, W. R., and H. E. Conklin. 1975. New farmland preservation programs in New York. Journal of the American Institute of Planners 41(6):390-396.

Clawson, M. 1962. Urban sprawl and speculation in suburban land. Land Economics 38(May):99-111.

———. 1972. America's land and its uses. Johns Hopkins Press, Baltimore.

Coughlin, R. E., and T. Plaut. 1978. Less-than-fee acquisition for the preservation of open space: Does it work? Journal of the American Institute of Planners 44(4):452-462.

Coughlin, R. E. et al. 1981. The protection of farmland: A reference guidebook for state and local governments. Report of the National Agricultural Lands Study. U.S. Government Printing Office, Washington, D.C.

Council on Environmental Quality [CEQ]. 1974. Environmental quality: Fifth annual report. U.S. Government Printing Office, Washington, D.C.

Dederiksen, R. I., A. R. Hidlebaugh, and K. O. Schmude. 1977. Potential cropland study. Statistical Bulletin No. 578. Soil Conservation Service, USDA, Washington, D.C.

Downing, K. B., K. P. Connaughton, and W. S. Hopkins. 1976. Urban owners of forests in western Oregon. Forestry Research Paper No. 26. Oregon State School of Forestry, Corvallis.

Esseks, J. D. 1979. Non-urban competition for farmland. In: M. Schnepf, ed., Farmland, food and the future, pp. 49-65. Soil Conservation Society of America, Ankeny, Iowa.

Fuguitt, G. V., and P. R. Voss. 1979. Growth and change in rural America. Urban Land Institute, Washington, D.C.

Gleeson, M. E. et al. 1975. Urban growth management systems: An evaluation of policy-related research, P.A.S. Report No. 309-310. American Society of Planning Officials, Chicago.

Gould, E. M., Jr. 1978. Wanted: High-satisfaction forestry. Journal of Forestry 76(11):715-718.

Gustafson, G. C., and L. T. Wallace. 1975. Differential assessment as land use policy: The California case. Journal of the American Institute of Planners 41(6):379-389.

Gustafson, G. C., L. Daniels, and R. P. Shirack. 1982. The Oregon land use act: Implications for farmland and open space protection. Journal of the American Planning Association 48(3).

Hady, T., and A. G. Sibold. 1974. State programs for the differential assessment of farm and open space land. Agricultural Economics Report No. 256. USDA Economic Research Service, Washington, D.C.

Hagman, D. C. 1973. Public planning and control of urban and land use development. West Publishing Co., St. Paul, Minnesota.

Healy, R. G., and J. L. Short. 1981. The market for rural land: Trends, issues, policies. The Conservation Foundation, Washington, D.C.

Kaiser, E. J., and S. F. Weiss. 1970. Public policy and the residential development process. Journal of the American Institute of Planners 36(1):30-37.

Koss, W., and B. D. Scott. 1978. A profile of western Washington's nonindustrial forest landowners. Department of Natural Resources, Olympia, Washington.

Lamire, R. A. 1979. Creative land development: Bridge to the future. Houghton Mifflin, Boston.

Lindeman, B. 1976. Anatomy of land speculation. Journal of the American Institute of Planners 42(2):142-152.

McBride, G. A., and M. Clawson. 1970. Negotiation and land conversion. Journal of the American Institute of Planners 36(1):22-29.

Miner, D. 1976. Farmland retention in the Washington metropolitan area. Metropolitan Washington Council of Governments.

Plaut, T. 1976. The effects of urbanization on the loss of farmland at the rural-urban fringe: A national and regional perspective. Regional Science Research Institute Discussion Paper Series No. 94.

Real Estate Research Corporation. 1974. The costs of sprawl: Environmental and economic costs of alternative residential development patterns at the urban fringe. U.S. Government Printing Office, Washington, D.C.

So, F. S., D. R. Mosena, and F. S. Bangs, Jr. 1973. Planned unit development ordinances. P.A.S. Report No. 291. American Society of Planning Officials, Chicago.

Stanford Environmental Law Society. 1973. A handbook for controlling local growth. Stanford University Law School, Stanford, California.

Thurow, C., W. Toner, and D. Erley. 1975. Performance controls for sensitive lands. PAS Report No. 307-308. American Society of Planning Officials, Chicago.

Toner, W. 1978. Saving farms and farmlands: A community guide. PAS Report No. 333. American Society of Planning Officials, Chicago.

Washington State Association of Prosecuting Attorneys. 1980. The county role in forest practice regulation in Washington. n.p.

Whitney, S. 1977. California's new timber laws. Cry California 12(2):13-17.

21

CLOSING THE AMERICAN LAND FRONTIER: COMMENTS

Robert G. Healy

At least two contributors to this symposium have referred to the longstanding American tradition of the open frontier. As many will recall from studying history, the census of 1890 revealed the closing of the national settlement frontier. And three years later, speaking to a group of historians gathered in Chicago, Frederick Jackson Turner presented his famous analysis of the role of the frontier in forming a unique American character, one very different from the European in its optimism, its expansionism, its egalitarianism, and its attitude about natural resources.

But in many ways the American land frontier did not close in 1890. Even as Turner spoke, a technological revolution in U.S. agriculture was causing phenomenal improvements in productivity, greatly relieving the pressure of population growth on the farmland base. A great rural-to-urban migration was just beginning which would over the next seventy-five years literally empty out the American countryside. And loggers had decades of old-growth timber left to cut in the Lake States and the South and Appalachia, not to mention the Northwest, where the process is still going on.

I believe that the closing of the American land frontier has occurred much more recently than 1890, so recently in fact that we are still reacting to it. As lately as 1960, reasonably fertile land could be purchased in many parts of the country for as little as $50 per acre. Rural land use changes, such as land clearing for agriculture or the purchase of southern land by the timber industry, could be made quite easily, because no other uses were making much claim on the desired land. Rural growth was welcomed and sought after. The 'last settler syndrome' discussed by Robert Lee is a creation of the 1970s, not before. If I had to pick a single landmark to date it, I would probably think of the popular vote in 1972 in Colorado to reject the Winter Olympic Games on the grounds that the state was already growing fast enough.

By a closing land frontier, I do not mean to imply a land shortage. But I do mean land scarcity. Land scarcity means multiple bidders for the same land. It also means that more land is being actively and intensively managed, with increased amounts of capital invested in changing various characteristics of the land. And it means that each use and each user will have more neighbors and more active neighbors than they have ever before experienced.

Now how can a society react to a closing land frontier? One way is to try to break the land constraint through the application of new technology. This is the essence of the move toward intensification and geographic concentration in forest management. Such is the state of existing forest management that we do not really need technological change, just wider application of what we know already. And if there is technological change—say in forest genetics—we can release millions of additional acres for competing uses. Now I did raise some caution flags in my earlier chapter. We need to make sure that the quality of land available to forestry does not go down too drastically. And we need to make sure that owners have knowledge of economic forestry opportunities and that parcel sizes are sufficiently large for scale economies. But all in all, intensification is a good way of dealing with the land constraint for the nation as a whole. It works less well when regional considerations are brought into the picture. It represents a sort of shrinking away from neighboring uses, allowing them to claim timberland whenever they get too insistent. In a sense it represents forestry moving when the forest manager (as in the old tale of Daniel Boone) 'sees the smoke from the cabins.' This represents, I think, a particular problem for the Pacific Northwest. Although this region contains some of the highest site-class land in the country and some of the best opportunities for intensification, it also offers some of the greatest demands for rural settlement, for recreational use, and for scenic preservation. It is easy to envision a scenario in which forest output is maintained nationally, but declines significantly in the Northwest.

Another way to respond to a closing land frontier is to separate uses at a finer scale. This is the trend at present in agricultural land preservation. As Don Miller mentioned in the preceding chapter, it means the forest equivalent of New York's agricultural districts. It means urban limit lines such as those painfully being put in place in Oregon and those which maintain the agricultural landscape in parts of Western Europe. It might mean something akin to the right-to-farm laws that have recently been enacted in a dozen or more states. These laws provide that in agricultural areas, where farmers have been there first, subsequent nonfarm rural settlers have no legal right to bring nuisance complaints against traditional farm management practices. In effect, the new settlers are on warning that this is a farming area and they live there under the farmer's terms.

A third way of dealing with a closing land frontier is for users to learn to get along better with their neighbors. One element, as Dean Cromwell has pointed out, is education. This cuts two ways. New forest residents are going to have to accept that cutting trees is an agricultural practice, little different except in time scale from the harvesting of crops, not an act of ecological vandalism. On the other hand, I think it inevitable that forest managers, both public and private, who want to continue to operate in areas of substantial rural settlement will not be able to have the same freedom of action that they have had deep in the woods.

There may actually be a certain rough justice in this. The geographic spread of our urban civilization, or to use Ed Blakely's term, of an 'advanced industrial society,' brings negative externalities to the forest owner in the form of constraints on his profit-maximizing behavior. But that very same process brings positive externalities to the forest landowner in the form of potential development value for his land. This development value is a windfall to the owner in the sense that it is created by social action and is not due to any act of his own. The landowner may not wish to exercise that development value soon, or may even profess complete lack of interest in doing so, yet I have rarely seen such an owner, whether farmer or forester, who is willing to give up the prospect of development without substantial consideration.

This rough justice, in the form of social windfall and social limitations, probably works best in the case of the large forest landowner, including the timber firm. Over the coming decades, these firms will be sustaining high costs as a result of public regulation of their management practices; they will also be making substantial profits on those tracts they choose to develop and dispose of. I would hope that the industry will look to the total picture of these gains and losses, rather than considering each tract in isolation.

The problem becomes much tougher when one considers the owner of an individual tract of land. If the parcel is large enough, and has substantial value for both fiber production and development, the owner might consider an idea called the RPD or 'rural planned development.'

This means partial, generally clustered, development, with the remaining productive land being managed as a sort of productive amenity. There are already some specific examples of this idea. In Virginia, in farming country at the base of the Blue Ridge Mountains, a developer decided not to cut his 300-acre tract into the fifty 6-acre tracts permitted by local law, but instead to cluster the fifty units and preserve more than 200 acres as a working farm, owned by the residents' association and leased to a farmer.

In California, there have been several rural planned developments proposed that would involve a combination of clustered housing and large tracts of grazing land. In West Virginia, a group of Washington, D.C., residents, instead of buying individual large-acreage parcels for second homes, pooled their resources and bought a 2,000-acre mountainside. The houses occupy only a few carefully chosen acres; the rest is managed forest.

Another promising idea is the community land trust, by which open-ended nonprofit entities buy land or receive it by donation. Increasingly, these organizations are not just preserving land but using it for farming or forestry or even limited development. They are not likely to maximize commodity output, but they are maximizing the satisfaction of their membership. There are many other institutional alternatives, including the option, suggested by Jack Larsen, of requiring rural developments to buy and manage their own open space.

The end of a land frontier is an opportunity, not a problem. One has only to look to Western Europe, where productive land uses coexist with a dense and amenity-conscious population, to see that a vigorous competition for land may be a blessing in disguise.

22

A NEW ZONING STRATEGY FOR FOREST LANDS

Keith Dearborn

A great deal is expected of America's forest land—recreation, scenic open spaces, watershed protection, wildlife habitat, and fiber production. In addition, incompatible land uses such as residential development are either directly permitted or indirectly encouraged by treating forest land as a reserve area to absorb future urban pressures. Rarely are priorities established by local governments. If they are, it usually is at the expense of forest uses. The result is an inconsistent land use management system that is typically unfocused.

Bringing order and focus to local government forest land use management is absolutely essential. Forest lands represent the largest existing land use in western Washington and Oregon. The future use of forest land may be the key to maintaining a community's 'quality of life,' for it is clear from public opinion polls that the Pacific Northwesterners' view of 'quality' is directly dependent—at least visually—on accessible forest lands.

Once our forest lands are gone we lose not only a special aesthetic resource but also a unique natural resource. The forest lands west of the Cascades are some of the most productive softwood lands in the world, equivalent to class I farmland. They are, to a great extent, still owned by a few large landowners. For example, in King County, three private companies own close to 80 percent of the undeveloped privately owned developable land. This pattern holds for most western Washington counties. These large owners face a difficult question—one that is often faced by the western Washington farmer: why maintain fiber production when alternative land uses offer a greater return on investment? Thus an effective strategy designed to maintain forest land uses must relate to and positively reinforce the fiber production needs of these large landowners. The long-term success of local growth management programs depends on the success of this effort, for successful implementation of a growth management strategy depends on the long-term forest use decisions of these major owners.

This chapter evaluates a typical zoning strategy (illustration A) in terms of the criteria offered by Cromwell and Vaux, and then proposes a 'new' approach responsive both to the land use trends described by Healy and Miller and to the needs of the large forest landowners.

Dean Cromwell suggests in his chapter that any forest land management strategy must 'address people's value perceptions, must find ways to facilitate communication between conflicting views, and must provide ways to resolve differences.' Cromwell agrees with Vaux (1982) that management problems are not tied to a particular place, hence are not geographic but rather political in nature, reflecting different points of view on the social value of forests. Therefore, a forest land management strategy must function as an acceptable compromise of these perspectives. In short, to perform as Cromwell suggests, it cannot be one-sided.

FOREST LAND USE TRENDS

Miller, Healy, and Cromwell each describe recent patterns of land utilization to which any effective forest land management strategy must be responsive. First and foremost, forest land values have risen dramatically in the 1970s—from $126 per acre in 1971 to a high of $569 per acre in 1979 (Chance n.d.). In many areas the market value of forest land far exceeds the timber value. Given current market conditions, continued production of fiber is not always what the local property tax assessor would call the 'highest and best use.'

A second trend that has been observed is the breakup of large ownerships into smaller and smaller parcels. Often caused by the 5-acre lot phenomena, it is also initiated by forest owners as a means of protecting their development potential from well-intentioned efforts by local governments to preserve rural character through large-lot (10 to 80 acres) forest zoning. A third trend follows in tandem with the second: ownership fragmentation by nontraditional forest owners. Many of these new owners are land holders or investors rather than land users. Others are motivated by what Miller calls high-satisfaction rather than high-yield objectives. These two trends are rapidly depriving local governments of the historical ownership characteristic that offers the greatest land use management opportunity—large ownerships and few owners. Instead of one or two major owners to deal with, planners are now having to respond to several owners with diverse needs. The 1980 census provides the best illustration of these trends: after about a 160-year hiatus, rural areas are again growing more rapidly than the nation's urban communities.

These population trends are leading to forest land conversion. Rough estimates suggest that Washington State's forest land will be converted to nonforest use at a rate somewhere between 15,000 and 84,000 acres annually (Chance n.d.). Washington's commercial forests could shrink by as much as 4 million acres between 1970 and 2020, an area roughly equivalent to King and Snohomish counties combined.

This loss of forest land as a fiber resource is of little concern to Pacific Northwesterners today. However, Vaux (p. 44) emphasizes a most critical trend: our land base is becoming increasingly limited. Traditionally, urban pressures and demands have been responded to by committing portions of the public domain to exclusive recreational or cultural uses—wilderness areas, scenic areas, and national parks. No longer is this approach the course of least resistance. The day will soon be upon us when we find that the addition of more of these permanent open spaces will be at a socioeconomic cost we cannot afford. Resolution of forest land use decisions will then begin to take on the same popular criticality as the loss of farmland.

NEEDS OF THE LARGE FOREST OWNER

The large forest landowner who has lands at the urban/ forest interface must decide whether to sell and invest in the pine woods of the South, where the growing season is more rapid and the urban pressure less intense. Fortunately, this decision is not being made easily and is not based exclusively on return on investment. It is just as hard for a large forest owner to give up prime forest land as it was for the former King County farmer in the Green River Valley to sell his farm to a developer or land speculator. The forest industry is rooted in tradition and attached to the land.

To retain forest lands, local governments must understand and respond to the needs of the forest owner. Just as early farmland preservation efforts were naive and ineffectual until the specific needs of the farmer were identified and responded to, a similar effort is needed for the forest industry. Until then, forest land use management will not have any long-term vitality.

Most important to the forest landowner is the need to minimize points of contact with non-forest users. Growing and harvesting trees is not a 'clean' business. For the urbanite, exposed to it for the first time, industrial forestry can be environmentally harsh and aesthetically unpleasant. To operate efficiently, modern forestry must be buffered from other uses. This means the forest industry can be an articulate and vigorous advocate for strong subdivision regulations and cluster development options. Furthermore, beyond site-specific development practices, the forest landowner needs to have urban uses contained and lands already committed to urban activity used efficiently. The biggest enemy of the forest industry has traditionally been urban sprawl and more recently large-lot rural zoning.

Forest landowners also need land use regulations that function as nuisance abatement tools. Modern forest practices, consistent with environmental standards, must be encouraged. In short, forest landowners need to have their lands treated as a resource, not a holding zone or recreational greenbelt.

TRADITIONAL ZONING STRATEGIES

Local government forest land use management strategies have typically taken one of two forms. Most frequently, forest lands are classified in some type of holding zone or urban reserve. They are colored green for reserve or open space on comprehensive plan maps, and the plan text describes policy objectives in attractive terms. However, the implementing zoning can be, as in King County, a forest-recreation zone with a 1-acre minimum lot size. This strategy fails on all counts. It serves nobody and fuels the forest land use trends discussed above. Nor does it represent Cromwell's effective compromise.

A second and slightly more modern strategy is set forth in illustration A, a large-lot agriforest zone (see Appendix 1). It is fashioned on the premise that forestry is another form of agriculture and can be treated in a manner similar to the treatment of dairy and berry farms. It is also based on the assumption that residential development can be discouraged by a large minimum lot size. Unfortunately this premise is based on neither analysis nor experience.

New efforts in agricultural land preservation that go beyond SCS soil classification as a basis for identifying valuable farmland have caused local government to realize that uniform approaches to farmland preservation will not work. The needs of the dairy farmer are quite different from those of the farmer whose livelihood depends on berries. It is just as unfounded to lump timber production with food production. Unfortunately, resource land use management is more complex. While both types of land use need buffering from urban activities and protection from nuisance complaints, the western Washington farmer's needs cannot be comingled successfully with those of the modern industrial forester.

In turn, experience everywhere with large-lot zoning leads to only one conclusion: a large minimum lot size simply causes larger lots to be created. It does nothing to reduce demand, hence does not reduce urban pressures on forest lands. In fact, the strategy portrayed by illustration A creates a fiction, relying on an artificial management tool called lot size. The forest land use trend ultimately will erode this strategy, and local government will again be called on to address forest land use management questions they had assumed were resolved by previous zoning efforts.

A NEW RESOURCE BASED STRATEGY

A new land use strategy is needed—one based on resource rather than use management. The new strategy must abandon even the notion of lot size, for the concept is inconsistent with forest management. A suggested forest management zoning classification is set forth in illustration B (see Appendix 2).

The forest management (FM) zone has only one purpose, to protect and encourage the long-term commercial use of our forest resources. It cannot be indiscriminately applied to all forest lands. Careful inventory must be done to identify those lands with soil, topography, and climatic conditions that merit a long-term resource commitment. Other factors should also be considered in identifying eligible forest resource lands: the presence of sensitive lands readily damaged by forest practices, scenic and aesthetic character, and recreational and housing requirements. In short, the land so classified must have a realistic chance of functioning as a long-term forest resource, because the FM zone is an exclusive use classification: no other land uses are allowed unless the applicant can demonstrate that such use will not interfere with timber production.

The FM zone relies on parcel size rather than lot size as the basis for regulation. The suggested starting point is 40 acres, but the size ultimately established is immaterial, for it conveys no development potential. Rather than striking a compromise between development and forestry using lot size as the vehicle, the FM zone avoids the issue completely.

The FM zone imposes a minimal buffer requirement which may be expanded when such lands are adjacent to scenic corridors. Although unstated in the zoning text itself, the buffer burden must be imposed on adjacent parcels not classified FM. The basic notion is that FM zoned lands should function as a resource, and the timber producer must be able to exercise the full range of practices needed for commercial timber production. The burden is on the adjacent property and not on the forest producer. Signs should be placed on roadways warning visitors that they are entering a forest management district.

Perhaps the greatest concern to the forest landowner generated by exclusive zoning is the fear that once classified there is no way out—the commitment is permanent. This concern stems from a historic mistrust of local government. Rarely have local land use decisions reflected the interest of forest landowners over the needs of other users—users who often represent more votes at the next election.

The FM zone specifically allows for reclassification. The process would be that used for rezoning. But action is not discretionary. When a proposed use poses no material interference with any adjacent timber production use, and the land proposed for reclassification is removed from the designated or classified timber tax status, the reclassification must be approved. Obviously, more elaborate or detailed standards may be used. However, the basic notion is that conversion is within the control of the forest landowner based on performance standards rather than the desires or preference of the voting public.

COMPARISON OF STRATEGIES

The FM zone meets Cromwell's conditions by defining a process. Criteria must be used to identify lands to which the zone may be applied. To the extent nonnatural resource factors are relied on in the land inventory, some political compromising can and will result. But the basic compromise is inherent in the terms of the zone itself. The forest landowner must accept an exclusive zoning classification, and in return the general public must agree to allow the owner to opt out based on performance standards rather than preferences.

The FM zone is also responsive to the needs of the forest landowner. There is no incentive for the nontraditional forest landowner to own land classified as FM. What can he do with it? Furthermore, there is no incentive to reduce parcel size or fragment ownership, because a smaller parcel size gains nothing. In turn, points of contact are minimized by placing the burden of buffering on adjacent nonforest uses and requiring that nonforest uses meet the noninterference performance standard. All in all, the FM zone responds more effectively to trends and needs than traditional strategies do.

CONCLUSION

Vaux, in describing the conflicts inherent in the urban/forestry interface, concludes by observing: 'The significant point here is that all of these things are real political and economic facts which make the present conflicts over forest land use predictable and, indeed, quite natural.' Traditional local government zoning strategies do not realistically address these natural conflicts, and they do not seek an effective compromise. Most often they focus on a large lot size as the common political denominator, and this approach is conceptually inconsistent with timber production. Furthermore, it has no rational relation to the residential carrying capacity of the land resource. It is time for forest land use management to reflect the characteristics of the resource and not an artificial compromise that in the long run serves nobody's interests.

APPENDIX 1

Illustration A

AGRICULTURE-FOREST USE DISTRICT

The primary purpose of the agriforest (AF) district is to protect and encourage the use of agricultural and forest resources. It is designed to encourage the type of activity that preserves, enhances, and supports the opportunity for the production of food and fiber. It is the intent of this district that such activity will maintain a sufficiently large scale of endeavor to ensure the economic feasibility of necessary support services to the resource utilization sector. In addition, the nature of nonagricultural nonforest activity allowed in this district shall be of such intensity as to preclude excessive tax burdens as a result of urban encroachment and attendant public service costs, provide visual relief to urbanization, and protect sensitive natural systems. Toward these ends, division or diversion of productive lands for nonagricultural or nonforest purposes shall be discouraged.

Permitted Uses

1. Agriculture, poultry raising, or accessory uses
2. Growing and harvesting of forestry products and accessory uses
3. Single-family detached dwellings, including factory built and mobile/manufactured homes, and accessory uses
4. Utility transmission lines necessary to serve the needs of the residents of the county
5. Single-family density transfer housing
6. Nurseries and greenhouses
7. Portable sawmills

Conditional Uses

1. Public utility substations
2. Animal hospitals, commercial kennels, public stables and riding areas
3. Churches
4. Parks
5. Sanitary landfills
6. Commercial excavation and processing of soil, gravel, peat, and other natural deposits
7. Home industry

Prohibited Uses

1. Commercial uses
2. Industrial uses
3. Throughput utility transmission facilities
4. Disposal of hazardous wastes

Tract and Yard Requirements

1. Minimum tract size: 20 acres per dwelling unit
2. Minimum front yard:
   (a) 30 feet on major arterials or 70 feet from the center line, whichever is greater
   (b) 25 feet on secondary roads or 55 feet from the center line, whichever is greater
   (c) 40 feet on state highways or 55 feet from the center line, whichever is greater
3. Minimum rear yard: 30 feet
4. Minimum side yard: 10 feet

Maximum Height of Buildings

The maximum height of residences in the AF district shall not exceed 25 feet, and accessory buildings shall not exceed 15 feet. Height restrictions shall not apply to agricultural or farm buildings or those associated with the growth or harvest of fiber.

Minimum Improvement Standards

1. On-site sewage disposal
2. Approved water system
3. Roads, right of way, and so forth, as required by county land development standards

Access

Every lot or tract segregated for development shall front on a street dedicated for public use or on a private road easement improved to adopted county standards.

Compliance with County Rules and Regulations

All buildings, structures, and uses shall comply with building codes, land development standards, water and sewer requirements, and other applicable rules and regulations of the county and/or the state of Washington.

APPENDIX 2

Illustration B

FOREST MANAGEMENT DISTRICT

The primary purpose of the forest management (FM) district is to protect and encourage the long-time commercial use of forest resources. It is established to identify geographical areas where a combination of soil, topography, and climatic conditions allow high-yield management practices to be conducted in an efficient manner, and to maximize the productivity of land so classified, consistent with standards and conditions that may be required to comply with any applicable sensitive area overlay district. Secondary purposes of the FM district are to protect and provide recreational opportunities and scenic open space to the extent such uses are not inconsistent with the primary purposes of the district. To the fullest extent possible, permitted uses in the FM district shall be buffered from all other land uses.

Permitted Uses

The growing, harvesting, and managing of forest products or any forest crop, in accordance with the Washington Forest Practices Act and Regulations adopted pursuant thereto, including but not limited to timber, Christmas trees, nursery stock, and floral vegetation. All uses necessary to the production of forest products and/or the harvesting and primary processing of timber, including facilities to allow the reduction in bulk and/or to facilitate the transport to secondary processing centers, shall be allowed as assessory uses.

Conditional Uses

Recreational facilities, including parks, playgrounds, hunting and fishing preserves, campground, recreational vehicle parks, youth camps, boat launches, and shooting ranges, regardless of whether temporary or permanent structures, facilities, or other improvements are involved, may be approved only when it can be demonstrated by the applicant that no interference will occur with permitted uses.

Prohibited Uses

Any use or activity that can or has the potential to interfere with the efficient management or long-term commercial productivity of forest resources shall be prohibited.

Tract and Yard Requirements

1. *Parcel size.* The minimum parcel size to qualify for the FM district is 40 acres, provided that smaller parcels may be designated if all other requirements of the district are satisfied; and

(a) The parcel has qualified as a designated or classified forest for purposes of property tax assessment; and

(b) Uses allowed in adjacent and abutting districts can be conducted in a manner that minimizes potential material interference with permitted uses.

2. *Setback buffer.* Permitted uses shall maintain a 100-foot undisturbed setback buffer from property lines and public road rights of way, and such additional setbacks may be required to preserve viewsheds from public roads designated as scenic corridors.

Maximum Height of Buildings

None

Minimum Improvement Standards

None

Access

Every parcel designated FM shall abut an improved public road or gain access to such road by private road easement improved to adopted local government standards.

Compliance with Codes and Regulations

All buildings, structures, and uses shall comply with building codes, land development standards, water and waste disposal requirements, and other applicable adopted local government codes and regulations.

Conversion/Reclassification

Any parcel classified FM shall be considered for reclassification to another use district upon request of the parcel owner if the parcel is removed from the designated or classified forest property tax assessment category and when uses allowed in the proposed district will not interfere with permitted uses in any adjacent and abutting parcel classified FM.

REFERENCES

Chance, Donald. n.d. New technologies in forest land use planning. Washington Forest Protection Association, Olympia.

Vaux, H.J. 1982. Forestry's hotseat: The urban/forest interface. American Forests 88(5):37, 44-46.

23

MANAGING THE URBAN/FOREST INTERFACE: A VIEW FROM FOREST INDUSTRY

William A. Atkinson

Crown Zellerbach owns thousands of acres of forest land where urban pressures are severe. We continually grapple with the problem of trying to run a forest business in the midst of urban pressure. It's quite clear to me that much of what is going on in California ought to be considered part of the problem rather than a solution. Certainly that level of costly bureaucracy and regulation scares the forest industry everywhere else in this country. The 'California solution' addresses all concerns except the concerns and rights of the private property owner. If California is the solution to the problem of the urban/forest interface, then you don't need anybody from forest industry contributing to this symposium, because we are not going to be around. We just cannot live with that kind of interference in the efficient management of forest lands. The forest industry today is not the forest industry of the 1970s. We are experiencing a deep recession, unlike anything in the last fifty years. Permanent structural changes are taking place. People and firms that have been in the business for decades are dropping out. On top of this our markets are changing; because of freight rates and Canadian competition we on the West Coast have lost many of our traditional markets. This situation has caused people in the forest industry to make major changes in their plans. A rethinking is under way. Where do we want to be located? What markets do we want to be in? What manufacturing plants fit into our long-term strategy and what plants do not? What lands do we want to keep and manage for wood production? Companies that survive are going to be much leaner than they were, much more competitive and aggressive. The California level of control over forest management came about in an era of high profits—the 1970s. We could afford nonproductive people on the staff to deal with field inspections, government permits, plans, hearings, appeals, and all the rest. We could even afford costly, petty, and unnecessary requirements. You are going to see timber companies fighting against this sort of thing much more fiercely in the future. In our view it's a fight for survival as a profitable business.

In many parts of the nation the forest industry finds itself with extensive landholdings in areas of rapid urban expansion. We generally didn't plan it that way. It just happened. In Washington, for example, the lowlands were logged off first—areas around Puget Sound near the water. Those areas have now grown up to young-growth merchantable stands and are ready for harvest. But these lowlands are now the population centers of 20th-century Washington. So, the way in which the forest industry responds to land pressures is important, merely because we control so much strategically located land. Management of urban/forest interface lands can be strengthened if planners work *with* the forest industry, and do not frighten it away.

Four areas will be covered in this chapter: (1) forest management from an industrial perspective (Why do we do the things we do?), (2) practical problems of managing lands at the urban/forest interface, (3) urban/forest conflict as viewed from the industrial side (How do we make decisions on whether to sell out or attempt to manage?), and (4) land management strategies that allow greater harmony so that many forest values are preserved.

AN INDUSTRIAL PERSPECTIVE ON FOREST MANAGEMENT

Forest Management Objectives

Directing attention to specific management objectives for industrial forests, you can divide management objectives into three general categories: those concerned with profits (particularly long-term profits), those concerned with a sustained wood flow, and those that have something to do with maintaining a favorable business climate.

Long-term Profits. An apology isn't necessary for a company's trying to make a profit from forestry. We will not stay in business very long if we cannot make profits. The recent recession brings home the importance of jobs, and timber's contribution to county and state budgets. But we have to be able to spend money wisely if we are going to stay in business. We have to be efficient about the way we manage our forests, and that's a very important point.

The drive for profits also demands that we justify investments in forestry. We can't go out and spend money willy-nilly. We have to be able to show that we can maintain an acceptable profit.

Sustained Wood Flow. Forest industries need a steady flow of wood into their plants. You cannot start up a pulp mill and shut it down and start it up and shut it down like you might turn a faucet on and off. It's too costly. The same thing is true, to a lesser extent, of sawmills and plywood plants and all the other kinds of plants. Companies need a steady operation of their mills, a reasonable amount of wood coming from the forest every year, not highs and lows. In order to do this, obviously, companies must maintain control over wood supplies.

We worry about the loss of wood fiber as lands are withdrawn from commercial production, and traditional development within the urban/forest interface does withdraw land from wood production.

A Favorable Business Climate. This objective essentially says to obey the laws. There are numerous rules and regulations in this business, and companies must live with those laws. That doesn't mean, however, that the forest industry needs to yield to every real or imagined pressure group, as we seem to have done in California. In most parts of the country we still fight for our right to manage lands profitably.

These, in general, are the kinds of objectives that we have in the forest industry. You can sum up by saying that our goal is to manage our forests profitably, efficiently, and scientifically.

Management Activities

How are industrial forests managed? Figure 1 depicts a typical forest management regime as found in Washington and Oregon. The horizontal axis represents time, starting with today, year zero, and going out fifty years. On this line we can portray management activities taking place on a forest acre over time. Start out with an acre that has just been clearcut, and begin management by planting. Plantations frequently must be protected from animals to keep deer, elk, and

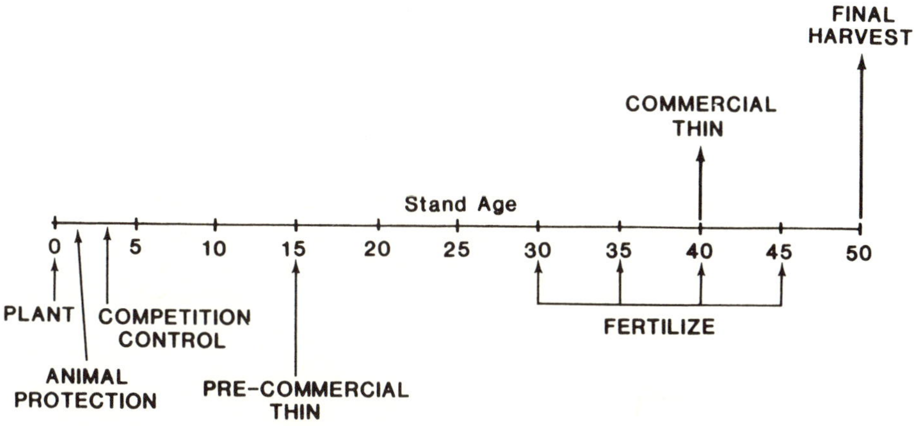

Figure 1. Example of a forest management regime.

mountain beavers from eating the seedlings. Brush comes in to compete with trees at age 3 or 4, and will probably have to be controlled. At age 15, we'll go into the young stand and do what is called a precommercial thinning to give trees room to grow. Fertilizer will be applied at ages 30, 35, 40, and 45. At age 40, we will finally begin to get money back by commercial thinning; that means we can sell the thinned logs. Final harvest will take place around age 50.

You should notice in figure 1 something that's very important about forestry. Most of the costs occur during the early part of the investment. Planting, competition control, and precommercial thinning are carried out fairly early in investment. You don't begin to get money out of these forestry investments until quite far along. This characteristic has a dramatic impact on the economics of forest management, and since profit is one of our major objectives, timing of costs and returns has a lot to do with the way forests are managed.

Forest management decisions also influence the quantity of wood produced. An intensive forest management program can produce more wood. Spend less money and you get less wood. This is not to say that intensive management is best. Maybe the cost of getting high yields is more than it's worth, and the optimal solution is somewhere in the middle. But there is a definite relation between inputs to forestry and outputs from forestry. You get more wood if you want to spend more to get it.

Characteristics of Forestry as a Business

There are several things that make forestry and timber production different from other lines of business. Two important ones are a long production period and high risk.

Long Production Period. How many businesses do you know take forty years or more to produce a salable product? Suppose we are trying to produce saw logs; a very small saw log is 12 feet long and 6 inches in diameter. But it takes at least forty years to produce trees that size. Depending on soil productivity it may take thirty years in some areas and sixty in others. Pulpwood for paper can be produced in twenty years or so—a shorter period because smaller trees are usable. But relative to most businesses, tree growing requires a long time. And this is very important. Because of a long production period the dominant forest management cost is interest on the investment.

If you had the opportunity of putting money away in a long-term bond at 10 percent interest, for example, or investing it in forestry, you must earn at least 10 percent on the forestry investment, or why bother?

Interest cost influences all decisions. For example, if we spend $100 on an acre to plant today at 10 percent, which is not an outlandish rate of interest, we must earn almost $12,000 from that acre in fifty years. People come to us and say, 'Hey, why don't you spend an extra $20 to $30 an acre to do something on your land today; it would make it a lot more attractive.' Maybe it would look prettier, but we have to get money back at the end of our investment period in order to justify it financially, which is the reason we're in business.

High Risk. Another characteristic that makes forestry a different kind of investment than many is the high risk involved. There are other businesses that have high risk, but we have high risk over a long period of time—fifty years maybe. There are many things that can destroy trees: insects, disease, snow/ice, windstorms, fire, animals, drought. There is a good deal of natural risk in forestry, and add to this the risk of high taxes and costly regulations. Why do we get into the business anyway if we can't earn a reasonable profit that allows a return for risk? You can put money into high grade bonds (which have practically no risk) and make a high rate of interest. We have to do better than that in forestry.

Competition for Capital

As foresters we must also compete within our companies for money to do reforestation, thinning, and other activities of intensive forestry. All companies have an internal financial hurdle, a rate of return that is required before funds are committed to projects. With rising interest rates it is becoming more and more difficult to justify forest management investments. Even if forestry manages to pass this financial hurdle, companies often have other and more profitable uses for their cash. Many expenditures are required just to stay in business: supplies, utilities, maintenance of equipment, and employee wages. Cash is also needed for strategic expenditures to move the company into new markets, block up timber supplies, and expand into new businesses. Finally, cash is needed for improvements in manufacturing plants, fixing up antiquated paper machines, converting sawmills to cut metric sizes, and changing to coal- and wood-fired boilers to reduce oil and electricity consumption. Most large forest companies today have major cash commitments that have nothing to do with forestry.

What all this means is that we must run our forest management operations as efficiently as we can. There is no extra margin for 'California-type' frills—management by group consensus or plans that aren't worth the paper they are written on.

Efficiency in Forest Management

Concern for efficiency in forest management leads the forest industry to seek out economies of mass production and large scale. Clearcutting does that. The clearcut silvicultural system (figure 2) allows large areas to be managed as a single unit. We can harvest, treat the slash, reforest, apply herbicides, and thin over many acres at the same time. It can be done in a mass production way, with big equipment, helicopters, large crews, and orderly logistics.

Figure 2. The clearcut silvicultural system.

There is efficiency in size here. It allows the concentration of attention on what trees need at a particular time in their life cycle. Since all trees in the area are the same age, we can give them exactly what they need. A good analogy between clearcutting and other silvicultural systems is a large, integrated school with a different classroom for each grade compared with a one-room schoolhouse. With a one-room schoolhouse, you've got kids from grades 1 to 6. You try to give all of them what they need at the same time. With a school that has separate grades, you can efficiently give the kids in the second grade what second grade kids ought to get, and so forth. The same thing is true of a forest. For a fourteen-year-old forest, the trees all need the same treatment at the same time. It might be fertilizer, it might be thinning. Efficiency is the dominating influence.

PRACTICAL PROBLEMS OF MANAGING THE URBAN/FOREST INTERFACE

Given this economic thrust in industrial forest management, what happens when we run up against those people who move into the forest to live their own rural life-style? In short, our nice, neat system of efficient forestry through large-scale operations is in big trouble. And we, as a company, are in big trouble, because now it costs us $300 an acre to establish a plantation while back in the hinterlands it only costs $100 an acre. Here is just a brief listing of problems that end up costing us money:

1. *Property line surveys*. No longer can a forester run cutting lines. Now everything must be done by a licensed (and expensive) surveyor.
2. *Large clearcuts, especially along roads*. Leave a buffer strip, people say. We sometimes do and see it blow down in the next big wind. Another—opposite—problem is trees growing up to block the view. You can't win either way.
3. *Traffic*. Big trucks and family cars don't mix, let alone the dust and noise of logging traffic. Easements and rights of way become more of a problem as population builds up.
4. *Water source*. Why is it the first thing people do, when they settle in their forest 5 acres, is buy plastic pipe and run it up the creek onto timber company land? The big company won't mind. But wait until they want to log or spray. Then all hell breaks loose.

5. *Herbicides*. The better the land for trees, the better the land for brush. As the population builds up in an area, we find more and more concern with aerial herbicides. We must leave wide buffers or use expensive hand-cutting methods.
6. *Slash burns*. Houses and slash burns together cause problems. In fact, the mere possibility of burning up someone's home is enough to make us look for alternatives to burning.

A number of other problems exist, but suffice it to say that the urban/forest interface is an extremely costly place to practice forestry and an uncertain place to make long-term forest investments.

THE FOREST INDUSTRY VIEW OF THE URBAN/FOREST INTERFACE

So far this chapter has presented a background on how forest companies operate. There are real pressures to make a profit and to examine the cost effectiveness of our practices. Given this situation, what do we do? How do we make decisions? Since land values in the urban/forest interface are rising, and are usually much higher than wood-growing forest land values, the companies that own those lands may well have an opportunity rather than a problem. How this opportunity is viewed depends on the situation. On the one hand, this high-value forest land may not be strategic to the business. Possibly it is easily replaced elsewhere, or just plain not needed. And maybe the urban pressures are so great that no one would elect to manage those lands for timber. On the other hand, there is land that you really need because your pulp mill or sawmill depends on it for fiber, or maybe the land is close-in and a low-cost source of delivered wood. Or possibly the land is at a low elevation and allows for winter snow-free logging. Decisions as to how to respond to urban pressures are often not simple, but the ultimate criterion is maximizing long-run profits for the company.

If urban/forest interface lands truly are more valuable for real estate than they are for forestry, then companies are going to be looking for ways to get the extra value out of the land. One way is to sell it, take the money, and do something else. Another way is to manage the land so that most real estate values are captured. But it's not realistic to expect timber companies to sit on high-value lands for very long, and continue to manage them for forestry.

What can be done to encourage the forest industry to hold on to its land and manage it to preserve forest values? The final portion of this chapter addresses this question.

STRATEGIES FOR INDUSTRIAL LAND MANAGEMENT WITHIN THE URBAN/FOREST INTERFACE

Consider a statement in Dean Cromwell's chapter, when citing Henry Vaux, to the effect that the urban/forest interface does not represent a geographical phenomenon, because problems are not confined to a particular place. Now if Vaux meant by this that we have pressures everywhere—people telling us how or how not to manage our lands—then I would certainly agree. But one must also point out that these pressures vary a great deal. There does exist a zone of intensive urban/forest conflict. Most land managers could draw it on a map. Management within this zone could very well be different from management in more remote areas. The forest industry is concerned that restrictions on the management of private property that are imposed to deal with a set of urban pressure problems in a specific area might be imposed everywhere. That worries us. The old foot in the door. Give in here and the same thing will be shoved down your throat on the other side of the mountain. Flexibility is the key. It's not beyond imagination that the forest industry might go along with specific restrictions—might even take the lead—if we knew ahead of time that these same restrictions would not be required where they are not needed.

So, there is such a thing as an urban/forest problem area, and a first step toward solution is to recognize that these areas and these problems are special.

Strategies for Harmony

Turning to some strategies for harmony between land uses in urban/forest interface zones, start out by assuming that it is best to keep the forest industry involved. It would be easier to deal with a few large companies rather than hundreds of small landowners. The question is how to allow the forest industry to extract some of the real estate values from its lands and yet not cause the companies to sell out and give up the whole thing.

I can think of two strategies, and they require innovation, in both planning and technical forestry. The first strategy concerns the pattern of development of forest land. The second is built around modifications of forest management activities that can go on in these zones.

Pattern Development

Remember that forest industries need reasonably good-sized blocks of land to manage efficiently. It is not possible to manage 40 acres here and 40 acres there; you need at least several hundred acres in a block. By allowing the forest industry to develop the small portions of their property that have high value for real estate, and keeping this development reasonably compact, you could then leave the rest of the forest alone to be managed for timber. There doesn't seem to be any basic conflict here. The forest owner wants to develop the same piece of ground most urban residents want to see developed. There are reasons why land is valuable for development—reasons that have to do with location, topography, view, access, water frontage, ease of utility, and so forth. The marketplace really guides the value of land for development.

And the converse is also true: the land needed for commercial forestry is not coveted by large segments of the urban population. They don't really want that land. Most people would be happy to see it left in forest production.

This may seem like a simplistic approach, but it makes sense and provides efficiencies for both the developer and the forester. Start out by looking at the land very carefully: inventory what you have; see where development makes sense, where the soils and topography are right, and where water, utilities, and roads can be brought in easily. Go ahead and let the landowner develop that land, or sell it to somebody to develop, but put strong restrictions on where development can take place.

You could also require a buffer zone around the developed area, and actually manage this zone as a buffer with fire breaks and screening so that the forest operation doesn't interfere with people living in the development. Outside the developed area and the buffer, the timber company would be free to carry on its own efficient forest management.

Modification of Forest Management

The other strategy for dealing with pressures in the urban/forest interface involves modification of forest management activities. We, in industry, are reluctant to give in on modifications of forest practices because of a fear that they will be forced on us everywhere. There must be an understanding that special situations do exist, and they may require special practices, but these practices are not needed everywhere.

For example, there are several silvicultural systems that can be used to harvest timber. You don't need to clearcut. We do it because it is efficient, but there are other systems. Figure 3 illustrates the seed tree system, for example. Here seed is dispersed from trees left scattered around the area. We may leave six, ten, or more seed trees per acre. They cast seed, and small trees start to grow. When new trees are established, the seed trees are removed with logging equipment. The seed tree system probably won't work too well with Douglas-fir, because this species doesn't produce frequent seed crops, but it does work for other species—pine or hemlock, for example.

Another system is called group selection (figure 4). Under group selection, small groups of trees are harvested, thus avoiding large clearcuts. The area always has the appearance of a forest, with small openings regenerating from natural seed or planting. This is a good system for properties where you want to maintain a continuous forest cover. It's not a very efficient system, because it doesn't produce as much wood as some of the other systems, but it has a place.

Yet another silvicultural system is called shelterwood (figure 5). Shelterwood is similar to the seed tree system except more trees are left—enough to provide shelter from hot sunlight and other harsh environmental factors for little trees coming up underneath. Eventually the overstory is removed (in two or three cuts), and this slowly releases the small trees. Using this system, trees are always growing on the land. Shelterwood systems work very well with shade tolerant species such as western hemlock.

It is also possible to tailor the clearcut system to blend in better with forest surroundings by using smaller units or more free-flowing and natural looking units. Much efficiency is lost by doing this, but it too may have a place.

Some other forest practices might be modified in the urban/forest interface: site preparation could be accomplished by clean logging or tractor scarification, rather than slash burns. Planting could be done with large stock that 'green up' the area quickly. Chemicals could be hand applied in sensitive areas, with buffers left along streams, and water quality monitored closely. Small, less destructive logging equipment and minimum impact harvest systems also are available. There are any number of technically possible forest management options that might be used in special circumstances. Since these practices all add to the cost of land management, we need to be sure that they are really needed, and that the owner has an incentive to carry them out.

CONCLUSION

It is technically possible to manage urban/forest interface zones so that people and trees exist in harmony. A change in attitude will be required on the part of everyone, as well as a good deal of imagination. I feel the forest industry ought to be involved and actually has a lot to contribute.

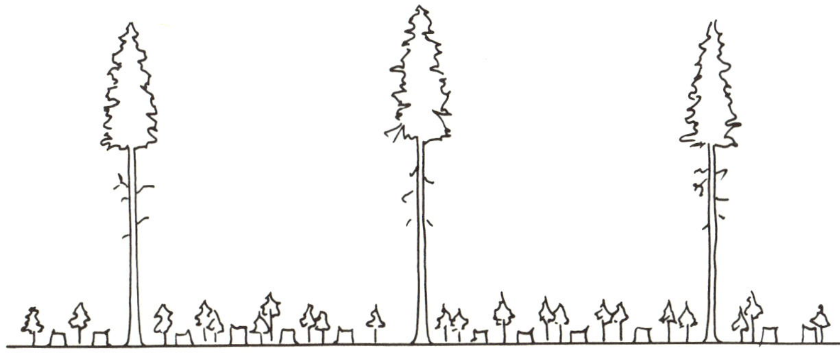

Figure 3. The seed tree silvicultural system.

Figure 4. The group selection silvicultural system.

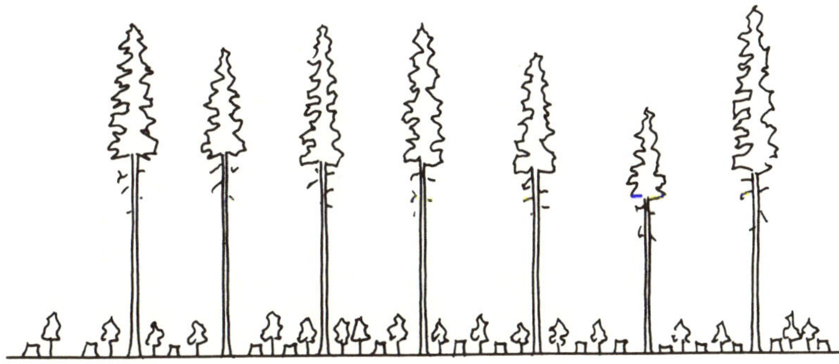

Figure 5. The shelterwood silvicultural system.

24

THE URBAN/FOREST INTERFACE: SUMMARY OBSERVATIONS AND RESEARCH NEEDS

Marion Clawson

The conference from which the chapters in this book resulted was an excellent one: the urban/forest subject is an important and timely topic, the conference was well planned, and the papers fit very well together.

The purpose of this concluding chapter is not to summarize or criticize the earlier chapters, but rather to point to some relationships among the separate chapters or topics and perhaps to suggest some overlooked or inadequately emphasized points, including some which seem not to fall clearly in any single chapter.

PEOPLE, TREES, AND FORESTS IN A CONTINUUM

People, trees, and forests exhibit a continuum: from the center of a large city where there are many people, few if any trees, and no forests; to the suburbs where many people live, where there are many trees but no forests in the sense of large groups of trees; to the urban/forest interface where urban uses are penetrating and intermingling with the forests; and to the strictly rural forests where there are few people. As with any continuum, there are many gradations from one situation to another. It seems useful, however, to recognize three major focal points or situations: strictly urban forest, urban/forest interfaces, and strictly rural forests. Each presents its own problems and opportunities—silvicultural, economic, and political or public relations. A brief consideration of the major aspects of these problems and opportunities may better set the stage for a consideration of the urban/forest interface.

In urban and suburban forestry, in the area where many people live and work, the first and most essential step is to secure the cooperation of people resident in the area or people using the streets, sidewalks, and parking lots where it is proposed to plant and grow trees. Unless their cooperation is secured and is manifested in the forestry program, failure is inevitable. Perhaps cooperation is most needed from youth—the street gangs. Without their cooperation, the life expectancy of the planted tree is very short. It is, of course, not enough to plant trees in an urban and suburban setting; they must be cared for throughout their life, and at relatively frequent intervals pruned, fertilized, sprayed, and otherwise taken care of. The emphasis in the urban and suburban areas is on the individual tree; the forest, as a group of trees with ecological interrelation, is missing or infrequent. The value of an individual tree is vastly greater than in the more rural settings, hence far more effort and money can be spent economically to manage an urban tree. At some date, when the tree is 'ripe,' it must be removed, for the safety of residents and passers-by, and for the health of other trees and shrubs on which it may encroach.

The date at which the urban and suburban tree is 'ripe' may not coincide with the age of maximum mean annual increment of wood growth or with the age of financial maturity. Indeed, one of the major problems of urban and suburban forestry is for the professional worker to decide when the tree is ripe and to convince people of the area that his (or her) decision is optimal from their viewpoint. A tree standing well past its prime becomes a major hazard to people. When the tree is ripe and is harvested, there is every reason to use the wood as economically as possible—for sawlogs, or peeler logs, if suitable for these uses, or as fuel. One of the ways of developing citizen cooperation in good urban and suburban forestry is to make use of the wood in ways that benefit local people—giving them an opportunity to buy firewood at favorable prices or plowing back into new plantings the money received from sale of wood.

I have been impressed and heartened at the progress made in urban and suburban forestry in recent years. Many organizations, public and private, have pushed urban and suburban forestry, and their efforts are beginning to pay large dividends. The American Forestry Association at its 1982 annual conference concentrated on urban forestry. Many who attended were young, many were women, and many had backgrounds in urban planning or disciplines other than forestry. I was particularly impressed with the recognition that trees must be planted and cared for, but also must be removed at some date. The naive preservationism—the concept that individual trees can be preserved 'forever'—which was all too common a few years ago is not in evidence. Most people like trees, are willing to spend money or to exert effort to have trees, and are increasingly becoming realistic if not expert in their management.

In heavily forested regions surrounding large cities, of which Seattle and Portland are prime examples, people are moving into surrounding forests to live, do business, and recreate, in a manner long common with similar population movements into agricultural areas. In some regions, the forests thus encroached upon are not highly productive in a wood growing sense, but in the Pacific Northwest the forests surrounding metropolitan areas are highly productive. It is not necessary to define the 'urban/forest interface' precisely; it is the general zone where significant penetration of people has occurred or is occurring but where there remains significant acreage of forest. It is a zone more than a line. The zone may be expected to be 'permanent' or at least to endure for several decades or it may be transitional, likely soon to move outward to a new location. Here, the individual tree is still important but the forest is more so. The silvicultural problems of stand establishment, stand maintenance and care, and harvest are important but somewhat different from those in the urban/suburban areas. In the interface zone, most of the surface of the land is covered by trees, but it is the clearings that give value to the whole area. The trees have significant cash values which can be captured at harvest time, but the land produces other values. Tree harvest is necessary in the interface zone, as it is everywhere, but conflicts may arise between harvesting and preservation. Such conflicts arouse most attention when they are between different landowners, but a landowner may find himself facing internal conflicts—whether to harvest an overmature stand for its significant income, whether to risk its blow-down, what to replant, how to manage the regeneration stage, and the like. Traditional forestry must undergo some modifications in the urban/forest interface, just as it must undergo modifications in the urban and suburban areas. The basic silvicultural, economic, and ecological principles still apply, but the circumstances—or the magnitude of the variables that enter into the formulae—are different, depending on the situation.

Foresters and forestry have traditionally been concerned primarily with rural areas, where the individual trees merged into and were subordinate to the forest as a collection of ecologically interrelated trees. While most foresters have conceded or proclaimed the values of the forest in addition to the growing of wood, it has been wood, in one of its many forms and uses, which has paid the bills and provided the tangible economic return to the forest owner. In the rural forest, trees must be reestablished after harvest, by natural seeding or by planting. While care during growth may be less intensive than is care for trees in an urban and suburban situation, some care—fire control, if nothing more—is essential. Harvesting presents both problems and great opportunities to manage the forest in new ways, if this is desired. Since the problems and opportunities of rural forestry have been the subject of most forestry education, research, and writing, they need not be explored further here.

LAND AREA AND OUTPUT FROM LAND

There is concern in many quarters these days about how the surface of the land is used. Less concern is expressed about the output from the land. 'Outputs' include all kinds of tangible commodities, such as wood, and all kinds of services, such as recreational opportunities and wilderness experience. With a growing population, the area of land per capita is shrinking; if output per unit of land area were constant, then total output and many kinds of outputs would also be shrinking on a per capita basis. But outputs per unit of land are a function not only of land input but of other inputs—labor, capital entrepreneurship, and, above all, technology. These inputs can be varied, and over time have varied greatly, with consequent major effects on outputs from the land.

In general, it is the land outputs that we seek—the time, the water, the wildlife, the recreation, the wilderness experience, and the beauty of the landscape. For some land uses the land area is more critical than output, but these are less common than the uses in which output is sought. Some persons occupy land for privacy, hence value land purely for its area: some buy land for speculation, and again it is land area that interests them. One can conceive of other situations where output from the land is relatively unimportant compared with land area. At the conference at which the chapters in this book were first presented as papers, there was somewhat more emphasis on use of the surface of the land than perhaps was ideal.

As a matter of fact, outputs from relatively constant acreages of land have increased greatly in the past generation or two. The agricultural story of the United States is well known: from a more or less constant crop acreage (declines at some periods offset by increases at others), with only one third of the labor force, and with about the same value of capital (but in different form), agricultural output has doubled. The persistent problem of American agriculture is surplus, not scarcity. In this respect, of course, the United States is extremely fortunate compared with many parts of the world. But the record of wood growth is equally spectacular. Since 1952, when the first modern forest inventory was taken, total annual wood growth has increased by more than 50 percent from a slightly smaller forest land area. If the best estimates of wood growth in 1920 are used as the base, the increase since that date has been more than threefold. Data on outputs of the forest other than wood are less complete and perhaps less accurate. In the national forests, as one ownership class of forest, the increases in recreation, wilderness use, watershed value, and wildlife have been as great as or greater than the increases in wood growth. There is considerable reason to believe that the increases in outputs from privately owned forests have also been great.

The explanation for both the agricultural and the forestry increases in output is the development and application of knowledge and new technologies, and the inputs of other factors of production in timely and effective ways. No knowledgeable person believes that all possible increases in output have been achieved; on the contrary, further substantial increases from the same or slightly reduced acreage are probable when known technology is applied more widely. There is every reason to expect the invention of new technologies that will have additional effects. These increases in output per acre are extremely important, for in them lies the potential for resolving many of the conflicts over land use.

The economist is likely to emphasize the role that increased prices or values of outputs play in inducing owners and managers to take those steps that in fact do lead to increased output per acre. In the case of forestry, higher stumpage prices have combined with many public programs, such as the control of wildfire, to help achieve much greater increases in wood production. There is nothing that so induces more intensive forestry as the prospect of a profit. Many a forest owner who now says he will never sell timber from his land will change his mind when a good opportunity presents itself, or his successor in ownership will sell the timber when a profitable opportunity arises.

The whole matter of supply response to price, for each kind of output, for each class of forest landowner, and for each forest situation deserves more careful research and consideration than it received at the conference. Supply analysis is notoriously difficult and the results are rarely above criticism from someone, but some conception of supply response is critical to all public and private programs involving forest land.

Forest outputs from the urban/forest interface are of several kinds, and realization of one kind of output is often in conflict with realization of other kinds. The family that establishes residence in an urban/forest interface zone values the beauty and appearance of the forest, the wildlife that lives there, and the general privacy provided by a home in a forest clearing. The family is likely to oppose the harvest of timber from the intermingled forest areas and also to oppose forest management practices such as spraying to control insects and brush and the burning of slash from harvested areas. The timber landowner who seeks cash income from his land wishes to harvest the trees at times and by methods of his choosing, and to apply forest management practices that increase his net income from the land. Some degree of conflict in objectives is inevitable. Similar comparisons could be made for other pairs of forest uses and outputs. Moreover, the conflicts are not confined to the tract of land in question, but spill over to adjacent lands as well. This leads directly to the matter of externalities, the subject of the following section.

EXTERNALITIES, INTERDEPENDENCE, AND PROPERTY RIGHTS

The concept of externalities is now widely understood, and the term 'externalities' is used frequently in popular as well as professional writing. Both concept and term appear in several of the earlier chapters. Briefly, one person takes actions which impose effects on other persons who were not a party to the decision. To the latter, the effects are an externality.

Most of the discussion and writing, especially the legal writing, focuses on negative externalities. The classic example is the factory that produces smoke harmful to neighbors while at the same time providing employment to workers and income to the owner. The landowner whose land use practices lead to silting of streams is another example of negative externality. Both law and economics have been much concerned with such negative externalities. Where do the primary actors'rights end and the rights of others begin? What is permissible, what is actionable in a legal sense, and what offends the general public to the extent that steps will be taken to reduce or eliminate the externality? When pressed, both economists and lawyers will admit that positive externalities exist also: when my neighbor has a beautiful garden, I benefit by the appearance of the neighborhood and by my visual enjoyment of his garden, although I have contributed nothing to its provision. This type of positive externality is particularly common and important in the urban/forest interface zone. Residents value the benefits of the surrounding forest yet may contribute little or nothing to the cost of providing those benefits. The beauty or amenity values of the forest may be high to the resident or the observer, yet the costs are all borne by the forest owner. Neither law nor economics has dealt much with these positive externalities; indeed, they are often difficult to measure, it is hard to see how they could be required by law, and yet they are often the circumstances which give value to the land receiving such positive external effects.

I prefer the term 'interdependencies' to 'externalities.' The term 'externalities' implies that the value flows, negative and positive, are all one way, from the decision maker to the recipient. The term 'interdependencies' recognizes that the flow is a two-way matter—that the recipient often takes actions which affect the first decision maker. To take the earlier case of the neighbor's garden: not only do I gain from his actions in providing a beautiful garden, but he gains also if I provide a beautiful garden. This reciprocal gain is the basis of many restrictive covenants in some residential districts, whereby all property owners agree to maintain a particular architectural style. Or it can be the basis for weed, pest, or disease control by each landowner on his property for the benefit of the whole community. The value of private property nearly everywhere depends just as much on what happens on adjoining or neighboring property as it does on what happens on the particular plot.

The private competitive market does not recognize externalities, negative or positive. The decision maker acts on the basis of the costs to him and the values that accrue to him. Except for being altruistic and civic-minded, he cannot take into consideration the costs he imposes or the values he confers on others. Some relief, but in practice not effective relief, can be secured against negative externalities by private lawsuits, citing the specific or common law against nuisances. It is often difficult to get courts to recognize such costs; it may be difficult to prove that specific actions have resulted in specific external costs; and the suits are expensive and the outcome uncertain.

Because the competitive market cannot recognize negative and positive externalities and because the private lawsuit is a poor remedy, attention has turned to planning and public control over private land use. In the history of the United States, the pendulum between private and public decision making over land use has swung widely. In colonial and early national days there were a good many public controls over private land, particularly as to its inheritance. These were gradually removed, and by the middle and late 19th century we had swung completely to private controls—fee simple ownership, from the center of the earth to the zenith of the sky. What the owner did with his land was his business and no one else's. To a degree not found in any other country, the United States gave the owner of the surface of the land the legal rights to subsurface minerals. From that extreme position, this country has gradually adopted many public controls over private land use. Many informed observers do not realize that such controls began with rural land and have been as stringent and effective there as the controls over urban land use. Today there is little or no land in the United States the use of which is not to some degree controlled by public action—zoning, subdivison, building code, health code, and other restrictions in the cities, but air and water pollution controls in the rural land uses. Although I have spoken of a swinging pendulum, there is grave doubt that the pendulum will ever swing back, away from public controls and toward freer private action on land use in the United States. It is not difficult to provide a strong case that public controls have been only partly effective and that sometimes they have been counterproductive to the ends sought by their advocates. But the controls do exist, they are likely to prove permanent in some form, and the individual landowner must take account of them.

In no country of the world is the use of all land determined only in the competitive market.

Land use controls are more common and more stringent in urban than in rural areas in many countries. Some of the controls over private decisions are by some unit of government, but some are tribal, or customary, or imposed by providers of credit. On a worldwide scale, the United States is the citadel of free private action in land use as well as in many other aspects of economic life.

The proliferation of public controls over private land use and the increased stringency of many such controls have substantially affected the meaning and content of private property ownership. The discussion at the conference made it clear that substantial ambiguities and uncertainties exist as to what rights the 'owner' has and what rights other members of the public have in a particular piece of land. What does 'ownership' mean when the 'owner' is significantly restricted in the use he can make of his land? If others have the right or power to limit the use of particular pieces of private land, just who are these other persons, where do they live, and what rights do they have? If someone has the power to control land use, then that person has property rights also; the value of the property to the 'owner' has been in some degree affected. Many owners will feel that the value of their land has been diminished because their freedom of action has been reduced; but a case can be made, at least in some circumstances, that values of land in the whole area have been increased because the public control recognizes the interdependencies among land parcels, and that, although the individual's rights have been limited, the overall effect is increased property values for everyone.

The effects of public regulations on private land use are not, of course, confined to land in the urban/forest interface zone but apply to much other land as well.

'Property' connotes value; if property rights are modified, one would expect that property values would also be modified. Economists have given only limited attention to the effects of control on land values, and to the distribution among persons or groups of the losses or gains in value. Lawyers have done relatively little to define the rights of others than the 'owners,' and, as noted, the effects of public action on the rights of the owners are somewhat ambiguous. The problem is only partly economic and legal; in part it is attitudinal and political. If people are dissatisfied with land uses, they are likely to seek through public action to modify those uses, and in the process to affect the rights and the values of the owners. The uncertainties of the political process, including the judicial process by which the laws are defined and sometimes extended, may detract from the values both to the nominal owner and to the public seeking the change in land use. If the latter have some power over use of private land owned by others, they too have some measure of 'property' right—often an ill-defined amount and kind.

This conference provided additional evidence that the whole subject of property rights, for owners and for the general public, requires more thought, more research, and more discussion than it has received heretofore—and more than it received at this conference.

FORESTRY PROBLEMS AND VALUES AT THE URBAN/FOREST INTERFACE

Earlier chapters and discussion at the conference made it clear that forestry at the urban/forest interface has special problems and special values. Tree harvest, stand regeneration, stand care, and ultimate harvest are a cycle, not invariable in form or in length but generally recognized as an appropriate way to consider forestry problems and values. In the urban/forest interface zone, there are often special problems in tree harvesting: people object to having the trees cut down, partly for aesthetic reasons but also on other grounds, or to the noise and confusion of the harvesting and log transportation process, or to dust along the roads, or to the size of clearcuts, or to some other aspect of harvesting. Yet if the forests are to produce the cash income to the forest owner and the economic values to the community of which they are capable, harvest at some date and in some way is absolutely essential. The arguments can range over when and how. Few people will oppose the reestablishment of the stand of trees, although there may be differences about what species to plant, the methods of planting or seeding, and other aspects of stand regeneration. There are especially likely to be disagreements about disposal of slash. The forest owner may prefer to burn it in order to clear the site for planting. The forest resident is likely to argue for removal of the slash, as chips or as firewood.

The new tree stand must be cared for and protected. While there may be disagreement on methods, there would generally be agreement that forest wildfires should be controlled or, as far as possible, eliminated. Controlled fire may be used as a conscious tool of management. On the subject of forest fire hazards, there are likely to be significant differences between the residents of the interface zone and the owners of the forested areas. The residents may seriously underestimate the fire hazard. No serious fires have occurred since their often brief occupancy of the area, hence in their view none are likely. Forest industry firms of long experience are acutely aware of the fire hazard, especially the risk of the highly infrequent but very severe fire. My limited observation of some of the urban/forest interface zones near Seattle is that the ingredients for a forest fire holocaust are rapidly building up.

There is likely to be disagreement and controversy over specific forest management practices from the time of replanting to the time of harvest. Shall the area be sprayed to prevent shrub growth or the growth of hardwoods? Each of these are serious threats in the Pacific Northwest if Douglas-fir or other conifers are sought for their ultimate timber values. On many sites well adapted to these species, the young seedlings have a hard time getting started and without some positive management practices may fail to get reestablished. But residents of the area may oppose such measures, on general environmental or naturalist grounds or because they fear externalities from the application of chemicals or for other reasons. The same applies to insect and disease controls. Some residents may oppose noncommercial thinning of stands, on aesthetic objections to the appearance of the trees cut but not removed. And commercial thinnings are likely to be opposed in the same way that final cuts are opposed.

There are also likely to be major differences in the values attributed to the noncommercial outputs of the forest—the recreational, wildlife, and amenity values generally. The forest owner naturally enough wants financial return from his land, and as a general rule this comes only from timber. Residents within the urban/forest interface zone seek other values. The forest landowners are generally not opposed to wildlife, watershed, and recreational values and as a matter of fact give such values more recognition than their purely monetary return would justify.

The foregoing problems are more acute in the urban/forest interface zone than in the rural forested areas, yet in their general kind they arise in forestry everywhere. Foresters must increasingly learn to work with people who are affected by decisions and actions in the forest, as the papers and the discussion at this conference repeatedly emphasized. There are few if any locations where the forest owner/manager today may reach decisions and take actions on the forest without consideration of the effect on people in and near the area. But the extent of these interactions with the public are greater, more acute, and likely to lead to more substantial modifications of the purely forestry decisions in the urban/forest interface zone than in more rural areas. Foresters simply have to learn new skills, as the discussion at the conference emphasized.

The discussion also made it clear that there are no simple, single, or easy solutions to these problems. On the contrary, decisions must be reached after extended and careful involvement of many persons and often will involve many small adjustments and compromises to reach an overall inclusive program of action. The Tiger Mountain case, discussed and described at the conference, is a good example of how prolonged and intensive discussions and negotiations lead to a series of actions that receive a grudging or an enthusiastic acceptance from most of the actors—a consensus sufficient to support action. Each other case or area must be handled in a way appropriate to its circumstances.

One matter raised briefly at the conference and in one of the chapters deserves more emphasis. Over the past generation there has appeared a large number of private consulting foresters who are able and eager to provide technical expertise to nonindustrial forest landowners—for a fee, of course. A generation or more ago, most foresters were in public employ; then the larger integrated forest industry firms acquired staffs of professionally trained foresters. But as recently as twenty years ago the owners of a small private forest typically had difficulty getting competent professional forestry advice and help. Today, any forest owner who is willing to pay for it can hire such professional expertise. Moreover, in many circumstances the gains in immediate financial returns will more than offset the costs. A nonindustrial private forest owner, whether large or small, today can buy or hire equipment such as a pickup truck or a power saw; but equally he can employ skilled professional forest management for timber harvesting and sale, replanting, and stand care and improvement. The availability of such professional help is largely responsible for the fact that annual timber growth per acre has increased more, absolutely and relatively, on the nonindustrial private forests of the country than has timber growth on any other major class of forest ownership, since 1952. My judgment is that the quality of forestry on nonindustrial private forests will continue to improve because of this factor and because the higher stumpage prices of recent years will make such improved management increasingly profitable.

SOCIETY'S STRATEGIES IN RELATION TO FORESTRY IN THE URBAN/FOREST INTERFACE ZONE

Because the competitive private market cannot take account of the externalities from forest management and because there exist substantial differences of objectives, if not outright conflicts, over forest land use and management in the urban/forest interface zone, society has resorted to a number of strategies and techniques to influence if not control the use of land and trees in this zone. These techniques include special laws, regulations under applicable law, purchase of development rights, special tax treatments, and others. These are considered in several of the earlier chapters and need not be repeated here. Miller has discussed such strategies in a comprehensive way; Cromwell has described in detail the programs in operation in California; and others have presented viewpoints on this general subject. Although a detailed review of these chapters and the discussion are unnecessary, some generalizations may be useful.

First of all, no strategy and no tool, no matter how skillfully crafted and used, can be fully effective in all situations and for all objectives. Strategies and tools of social control may be effective to some degree in some situations, but there is no 'all purpose, complete remedy' available now or likely in the future. Special tax treatment, to take one example, may influence land use—forested and other land—in some circumstances and for some periods of time, but experience in many states has shown that the effectiveness of this tool of land use control is severely limited at best.

A corollary is that strategies and tools for influencing or controlling the use of forest or other land are more effective in combination than singly. Each reinforces and strengthens the others. But the coordinated use of several strategies and tools is often difficult. For one reason, each may be in the hands of different groups or different agencies—the tax tool in the hands of the assessor and tax collector, the public purchase tool in the hands of a special agency, the control of public services in the hands of still another agency, and so on. Even with the best will in the world, the effective coordination of different groups or agencies presents difficulties, and agency defense of turf against what are perceived as raids on jurisdiction and power is surely not unknown. Nevertheless, coordination of different strategies and tools is surely an important objective.

Miller particularly, and other speakers to some extent, emphasized that the combination of strategies to control forest land use in the general urban/forest interface zone should take into account the particular circumstances of each area. The ratio of people to land in each major part of the interface zone will have a marked effect on the optimum combination of strategies, for instance. The rate of population growth and of development in each major part of the zone is another significant factor. Where the number of people resident in the area is relatively large or where the rate of development is relatively rapid, the need for public controls over land use and land development and the best tools for such control are different from those in areas where present settlement is sparse and development is slow. I personally found this to be true in suburban development districts generally, and it surely is in forested areas subject to residential and other development. Where there exists strong pressure for development, the process will be different from where there is little such pressure. One part of this difference is that the actors will be somewhat different and the monetary values involved will be greatly different.

All of this is to say that the economic forces of the marketplace must be recognized in any strategy to influence land use in the interface zone—even when the objective of the strategy is to oppose or reject the marketplace solution. Plans and the tools of the planner can affect land use and land development but only within some range of economic forces. When the profits from development are very great or the pressures for development are very strong, the influence that planners can have is less than where the profits are small or the pressures light. Groups attempting to influence or control land use and land development in the interface zone are more likely to be successful in attaining their objectives if they carefully and objectively examine the economic environment and economic forces in the area with which they are concerned.

NEED FOR RESEARCH

Every professional conference reveals areas or fields of knowledge in which more and better information and greater understanding are badly needed—in short, where more research is needed. The conference on the urban/forest interface was no exception. Researchers are often derided by nonresearchers because the one sure result of every research project is a call for more research—and for more funds to support research. Such criticism profoundly misunderstands and misrepresents the nature of the knowledge accumulation process.

Every piece of research reveals some measure of truth—if the research is well and innovatively planned and conducted, a great deal of new truth; if poorly planned or badly conducted, much less new truth. But, at best, the research reveals more ignorance than truth. The researcher climbs one ridge as he crosses the plains of knowledge, but when he reaches the crest of that ridge he finds new areas and new ridges ahead, stretching to a new horizon. The researcher would be seriously derelict in his responsibility if he did not provide others with his best judgment as to the most effective route to future knowledge. It is in that spirit that the following relatively few and tentative suggestions for directions of future research on forestry at the urban/forest interface are made.

When it comes to land use in general and to forest land use in particular, numerous questions arise. First of all, just how accurate are the available data on land use? Or, to put it differently, within what limits and for what purposes are the data sufficiently accurate to sustain the analyses we wish to make of them? There is no simple answer to such broad questions, but researchers and the public should always look closely at the origin, nature, and probable accuracy of available data, before they make uses of it that may be unsustainable. More particularly, what significant changes in 'gross' land use are concealed within the usual figures of net land use changes? There conceivably could be, and some of us believe there have been, major changes in use of particular parcels of land even while net land use changes were minuscule. Studies should be made tracing the history of particular tracts of land over considerable periods, to supplement the studies that measure land use at intervals. Have the shifts in land use between forestry and other uses lowered or raised the average productive quality of forest land? Healy, in his chapter, points to the conflicting evidence on this point.

What accounts for the confusing and annoying discrepancy in data on numbers of farm forest owners as reported by the Forest Service forest inventory and as reported by the Census of Agriculture, and for a similar discrepancy between acreage of farm forests as reported by the Forest Service and the acreage of farm woodlands as reported by the Census of Agriculture? As is often the case, highly aggregate figures, such as a national total, may seem reasonable and differences may seem explainable; but when state and county figures are examined, the discrepancies are large and explanations are less apparent. It is hard to believe that both sets of data are accurate. Which one is dependable and for what purposes?

Also in the forest land use field, what have been the real and clear effects of special tax laws, minimum tract size laws or regulations, and other measures designed to keep some forested areas in forest? There has been a good deal of discussion of this general matter but very little hard incisive research directed at forest land. What has been the response, over significantly long periods, of nonindustrial private forest owners to higher stumpage prices? How much has the greatly increased timber growth on nonindustrial private forests been due to higher stumpage prices and how much due to various public programs, such as fire control, subsidy payments, and the like? How has this worked out on specific tracts of forest land, in specific locations, and not merely in large aggregated areas or totals? What steps might realistically be taken to draw the smaller nonindustrial private forest owners more directly into the market for stumpage? Are, in fact, any special measures needed to draw into the market the owner who has a significant volume of timber to sell in an area where good markets exist?

Research is needed on land use laws and controls as they apply to forested areas. Perhaps the first issue is how to reconcile the interests and approaches of the lawyers, as exemplified by Huffman's chapter, with the interests and approaches to the same subject by the urban planners, economists, sociologists, and others dealing with these same problems. What forces and kinds of knowledge motivate legislators and judges in their actions on forest land use matters? Many lawyers and others will argue that the judges simply interpret the law, trying to decide what the legislators really had in mind by confusing, unclear, and inapplicable language. But both judges and legislators get their ideas and facts from somewhere; that these concepts and facts are often at variance with those of professionals does not deny that some source was used by the legislators and judges, even when they may have been unaware of the origin. In a complex world with myriads of governmental policy issues, how might legislators and judges be better informed on the best available professional knowledge? Surely, incisive research here would be invaluable.

The need for public involvement in planning and decision making on natural resource use generally and on forest use and management in particular is usually acknowledged and was emphasized by several speakers at the conference. But few persons familiar with how this process works today are wholly satisfied with it. How can the general public be better informed and more capable of making a rational judgment, in a world so full of policy issues and where the flow of 'noise' if not information is so extensive?

The urban/forest interface conference revealed the need to measure more accurately, more easily, and more quickly, and in a way the general public can understand, the nonmonetary values of forests. This is a continuing problem and perhaps no wholly satisfactory answer is possible, but it is also a subject to which more research attention might well be directed. It is a subject on which there has been a considerable body of research, of an increasingly sophisticated economic character, but it is doubtful if accuracy of results has increased, and surely generally understanding has not. What spatial mixture of timber production and other uses of a generally forested area would maximize values to all owners and occupants? Would values be greater if 100 residential families each had a 5-acre tract or if each had a 1-acre tract and the remaining 400 acres were held in a common, for use by everyone, and managed to provide maximum multiple use values? Most analyses of land uses and land values omit consideration of the spatial intermingling of different uses—the 'parcellation' and 'juxtaposition' problems discussed by Healy. Surely, some patterns of intermixture are economically preferable to others, and research should be able to measure the values involved.

The need to define more precisely the property rights of the owner and of nonowners seems to me inherent in the chapters by Healy, Huffman, and Hagenstein. It was agreed that some of these rights are indefinite and undetermined at present. Just what are the legal, economic, and political rights of the 'owner'? And what are those of the nonowner, whether adjacent or distant? Clearly, the latter has some rights to a tract of land if he is able to prevent the nominal owner from using the land as the latter would choose. But just how may these rights of the nonowner be defined, their value measured, and their extent limited in some degree? By what means can nonowners be made to respect more exactly the rights of the owner? How can trespass and other unauthorized uses be reduced or eliminated? There are indeed laws against trespass, but they are often ineffective. Trespass must be detected, proof must be obtained, and court action is often costly and the results uncertain. Trespass laws are effective primarily in discouraging would-be trespassers, not in punishing actual trespassers. Trespass is at least as much a sociological as a legal problem; the solution to extensive trespass must lie in the inculcation of different attitudes on the part of potential trespassers. Unless trespass can be effectively controlled, profitable management of forest land, whether in the urban/forest interface zone or elsewhere, is seriously threatened.

The silvicultural and management problems of forested areas within the general urban/forest interface zone are serious, as some of the earlier chapters describe them, and should yield to further research in such specific situations. The general principles of silviculture are valid everywhere, but their application must take account of wishes and attitudes of owners and others in the interface zone. Perhaps the most important and difficult problem is to establish in the minds of residential and other owners some sense of time and the changes it brings in every forest. The present stand of trees will change over time, irrespective of management and efforts to prevent change, but management can affect the rate and to some extend the direction of change. Change brings opportunities and costs. Time also includes risks—of insect infestations, disease, blowdown, and fire, particularly; and most of these risks manifest themselves at intervals, often at very long intervals, and not regularly year by year. The average user of natural resources grossly underestimates the potentials of these infrequent but severe hazards. One task of research is to find ways of bringing this understanding to the average occupant of a forested area.

Several chapters emphasize the variety of values flowing from a well-managed forest—the multiple use values, if you will. But the definition of these values, and their measurement—each as contrasted with the rhetoric that asserts them—present difficult problems for research. It is essential that foresters, landowners, planners, and others recognize that such values exist and must be planned for; but it is equally essential that they understand that such values are not infinite. A beautiful view of a forested scene is desirable and valuable, but its value is not infinite. The need is for research to define and measure these values realistically while at the same time not denying their existence. Beyond measuring the values involved, research might well be directed at the equity problems: who gains the values and who pays the inevitable costs? Such research is difficult and unlikely to yield results that will be above criticism; but it might well produce more accurate estimates than the unsupported assertions of values by persons who have made no effort to quantify the values they assert.

Good, imaginative, and hardheaded research can provide the information and concepts that can significantly raise the level of debate about specific issues in the urban/forest interface. But no amount of facts and analysis will solve policy problems. That comes out of the value systems of the participants as much as it does out of the objective facts of the situation. One must be prepared to defend research against both its critics and its naive supporters.

NEED FOR SOCIAL INVENTION IN THE URBAN/FOREST INTERFACE

The need for social invention in the interface is greater than the need for research. Research is concerned with ascertaining new facts, developing new concepts, and explaining relations among variables or forces. As such, it is invaluable to the formation of wise social policy. Invention seeks to develop wholly new ideas, often by substantial rearrangement of old ideas into new forms, and to provide a mechanism whereby new and different actions may be taken. Invention may rely on the results of research but it requires more than the usual research activity. Research that outlines alternative lines of action is approaching invention; invention that starts with an examination of past and present relationships is approaching research; there is not a clear and sharp line between them, but the emphasis is different.

A few examples may make clearer the concept of social invention. The whole concept of land use zoning, to permit certain land uses in some areas but not others, was a social invention. The idea of long-term, regularly amortized mortgages with fixed interest rates on residential property was a social invention that came out of the Great Depression. The idea of a profession of private consulting foresters, whose services are available for a fee to any interested forest landowner, was another social invention.

Before going on to suggest, very tentatively, some areas toward which social inventions in forestry in the urban/forest interface might most profitably be directed, a few generalizations about social inventions as a whole may be useful. First of all, inventions, whether social, mechanical, or biological, are rarely the work of one person alone. Others are almost certainly involved, and difficult problems arise in deciding who should get the credit for a particular invention. Failures along the invention trail are almost always encountered; sometimes they help to guide the direction of success. The name associated with a great invention of the past is nearly always that of the person who first made the invention into a sufficiently workable device or applicable idea, after which refinements could ensue. Not all inventions seem socially desirable in the light of subsequent history. Once an invention comes into general use, it becomes part of the social landscape, taken for granted as 'natural.' For instance, to take one of the earlier examples, most people in the United States a few years ago took as established fact the regularly amortized residential mortgage, at a fixed interest rate over a long period, not recognizing that this was a social invention of about a generation's duration. The virtual disappearance of this type of residential financing in recent years of high inflation has shocked many persons who had taken it for granted.

A useful social invention for land use and occupancy in the urban/forest interface zone would be new and clear concepts of property rights of landowners and nonowners. There is a good bit of experimentation going on, and partial or implicit concepts must exist to underlie some of the actions taken by land developers, property owners, legislators, and the courts. But a really comprehensive definition of such rights, methods to measure them and the values they create, and measures to induce more of the public to accept whatever rights are established would be highly constructive. While research would be helpful here, invention should go beyond present and past experience.

Social invention in forestry for the urban/forest interface zone could productively help to develop a new profession, that of the interface forest manager. On the one hand, he or she would be similar to the consulting forester who works in rural areas. On the other hand, he or she would have much in common with the landscape gardener of the more closely settled urban areas. Such a person should be informed and competent in the technical aspects of forestry, but ability, skill, and experience in working with people, knowledge of business including tax laws, and ability to organize and direct specific operations on single or intermingled tracts of land would also be necessary. My impression is that many persons have bought forested tracts in the interface zone with very naive ideas of how to manage their land—perhaps most dangerously of all, an idea that the forest needs no management, that the trees are able to take care of themselves unaided and indefinitely. Pioneers in urban/forest forestry may not have an easy time, professionally or economically—inventors and pioneers often have hard times—but there is a real service to be performed and a real need for a somewhat different profession than now exists.

Another field in which social invention in the urban/forest interface zone might be highly productive is the development of concepts of group or area management of land and associated resources. In particular, the problems of optimal spatial relationships and of externalities or interdependencies, previously discussed, might be solved by some sort of group approach to resource use and management in this zone. Experience elsewhere may provide some suggestions. Many suburban residential developers, especially in the East, have constructed dams and created artificial lakes, sacrificing to the lake area some potential building sites in the belief that the lake added more to the value of the remaining area than was lost by giving up some building sites. There has been a substantial experience with such lakes and some research about them. They are often useful in attracting buyers to the houses in the development, but have not infrequently become liabilities some years after the developer moved on. Owners of farmland have created irrigation, drainage, and soil conservation districts, to attain a greater or lesser degree of control over individual use of land in the interest of the group as a whole. These have been effective in varying degrees and have not been free of problems; but, again, they might offer some suggestions for the development of a group approach to land and forest management in the interface zone.

If the externalities and the opportunities for significant income from the forest are to be taken fully into account in the interface zone, then some management areas larger than individual landholdings may be necessary. This might well entail the creation of forest management districts of some kind. There would seem to be room here for some social invention. What form of area approach seems most likely to be successful? How could individual landowners be persuaded that it was in their interest to join? How might such districts or grouped enterprises best function? What would be the role of the 'interface forester'? Should various public programs seek to operate chiefly or wholly through such districts or area groupings? For instance, if there are public subsidy programs for tree planting or other aspects of forestry, should they be funneled through these districts? Should fire insurance on residences be conditioned on participation in area resource management including fire prevention on an area basis? These are only some of the questions that should be faced, first on a conceptual basis and later in actual practice.

These examples of possible social inventions for forestry in the urban/forest interface zone are more illustrative than all-encompassing. Although most social inventions are the work of more than one person, there is nearly always one person with a vision or an ambition or a driving energy without which the whole enterprise would never succeed. While any author or observer may make suggestions, no one can hope to include all possibilities. The great strength of the American society and economy is its multifaceted character—millions or thousands of persons, all seeking to develop their own ideas, some succeeding, many failing, and some of the successes of dubious social value. This characteristic of the economy and society as a whole applies fully to forestry in the urban/forest interface zone. About all that society as a whole and noninventors as a class can do is to applaud, help to finance, and reward social invention.

ABOUT THE CONTRIBUTORS

Glenn J. Amster is an attorney in the Seattle firm Hillis, Phillips, Cairncross, Clark and Martin and is Article and Book Review Editor for the *Washington University Law Quarterly*.

William A. Atkinson is the manager of Crown Zellerbach's Forest Research Division, Wilsonville, Oregon. Prior to joining Crown Zellerbach he served on the forestry faculties at the University of Washington in Seattle and the University of California, Berkeley.

Edward J. Blakely is Professor of City and Regional Planning at the University of California, Berkeley, and is the coauthor of *Policy Implications of California's Changing Life Styles* (1978) and *Resources of Recent Migrants to Rural Areas for Economic Development: Policy Implications* (1981).

Brian Boyle is Washington Commissioner of Public Lands and, by state statute, Administrator of Washington's Department of Natural Resources.

Gordon A. Bradley is Associate Professor of Forest Resources Planning at the College of Forest Resources, University of Washington, Seattle, Washington. His primary teaching and research interests include organizational response to integrated land management planning, forest land use planning and evaluation, and site planning and design.

Marion Clawson is currently a consultant to Resources for the Future, a nonprofit private research and educational institution financed by the Ford Foundation. He has written extensively on subjects related to forestry and he served as Director of the Bureau of Land Management, USDI, from 1948 to 1953. His publications include *Suburban Land Conversion in the United States: An Economic and Governmental Process* (1971), *America's Land and Its Uses* (1972), *Forest Policy for the Future—Conflict, Compromise, Consensus* (1974), and *Forests For Whom and For What?* (1975).

Dean A. Cromwell is the Executive Officer of the California State Board of Forestry, Sacramento, California. He previously served as Executive Secretary of the Western States Legislative Forestry Task Force.

Keith Dearborn is an attorney and land use planner in the Seattle firm Dearborn and Associates.

Perry R. Hagenstein is President of Resource Issues, Inc., a Wayland, Massachusetts firm specializing in natural resources policy, economics, and management. He served as Senior Policy Analyst, U.S. Public Land Law Review Commission, Washington, D.C.

Robert G. Healy is an economist and Senior Associate at the Conservation Foundation, where he directs the Rural Land Market Project. He has written extensively on land use issues and policies including books on state regulation of land use, the California Coastal Commission, and policies for national forest lands. His publications include *Land Use and the States* (1976) and *The Market for Rural Land: Trends, Issues, Policies* (1981).

James L. Huffman is Professor of Law and Director, Natural Resources Law Institute, Lewis and Clark Law School, Portland, Oregon.

Rachel Kaplan is Professor of Natural Resources and Doctoral Program in Urban and Regional Planning, and Associate Professor of Psychology, at the University of Michigan. She has written extensively on subjects related to landscape assessment, citizen participation in environmental design and planning, and the role of natural environments in psychological well-being. She is coeditor of *Humanscape* (1978) and coauthor of *Cognition and Environment* (1982).

Jack Larsen is Vice President, Regulation and Environmental Affairs, Weyerhaeuser Company, Tacoma, Washington. He previously served as Director of the Washington State Department of Commerce and Economic Development.

Robert G. Lee is Professor of Forest Sociology at the University of Washington, College of Forest Resources, Seattle, Washington. His teaching, research, and publications focus on forest resource sociology with emphasis on comparative cross-cultural analysis of forestry institution building.

Donald H. Miller is Professor and Chairman, Department of Urban Planning, University of Washington, Seattle, Washington. He has written extensively on plan and program evaluation and previously served as Chairman of the Seattle Planning Commission.

Richard L. Morrill is Professor and Chairman, Department of Geography, University of Washington, Seattle, Washington. His major research interests include location theory, urban and regional development, and population settlement and migration.

Clark Row is Economist, Forest Resource Economic Research Staff, USDA Forest Service, Washington, D.C.

Roger A. Sedjo is Senior Fellow and Director of the Forest Economics and Policy Program, Resources for the Future, Washington, D.C. He has written extensively on international trade in forest resources, multiple use management of forest resources, and the economics of nonindustrial private forests. His publications include *Policy Alternatives for Nonindustrial Private Forests* (1978) and *Governmental Interventions, Social Needs and the Management of U.S. Forests* (1983).

David M. Smith is Professor of Silviculture, School of Forestry and Environmental Studies, Yale University, New Haven, Connecticut. He is the author of *The Practice of Silviculture* (1962).

Daniel Syrdal is an attorney in the Seattle firm Syrdal, Danelo, Klein and Myre. His practice is concentrated in the area of natural resources law.

Richard D. Taber is Professor of Forest Zoology and Wildlife Science at the College of Forest Resources, University of Washington, Seattle, Washington. His major research interests include biology and the conservation of birds and mammals, and wildlife and human culture.

Byron R. Thomas is the Principal Soil Scientist, Bureau of Land Management, Department of Interior, Portland, Oregon. His primary research focus is mantle stability and soil compaction.

Thomas R. Waggener is Associate Dean for Academic Affairs and Professor, Forest Economics, College of Forest Resources, University of Washington, Seattle, Washington. His teaching and research focus is forest economics, policy, and regional impact analysis. He served as staff economist/analyst for the Resources and Policy Evaluation Group, Public Land Law Review Commission, Washington, D.C.

INDEX

The Geo. S. Long Publication Series

American Forest Policy in Development by Stephen H. Spurr

Man, Land, and the Forest Environment by Marion Clawson

Renewable Resource Management for Forestry and Agriculture edited by James S. Bethel and Martin A. Massengale

Tree Growth and Environmental Stresses by Theodore T. Kozlowski

World Trade in Forest Products edited by James S. Bethel

Land Use and Forest Resources in a Changing Environment: The Urban/Forest Interface edited by Gordon A. Bradley

DATE DUE
